ARCHAEOLOGY
AND
PHOTOGRAPHY

ARCHAEOLOGY AND PHOTOGRAPHY

Time, Objectivity and Archive

EDITED BY
LESLEY McFADYEN AND
DAN HICKS

Routledge
Taylor & Francis Group

LONDON AND NEW YORK

First published 2020 by Bloomsbury Academic

Published 2020 by Routledge
2 Park Square, Milton Park, Abingdon, Oxon OX14 4RN
605 Third Avenue, New York, NY 10017

First issued in paperback 2021

Routledge is an imprint of the Taylor & Francis Group, an informa business

First published in Great Britain 2020

Cover design by Irene Martinez Costa | Cover photograph: The shadow of archaeologist
Richard Atkinson, and his camera, taking a photo of an archaeological excavation
© Ashmolean Museum, University of Oxford

A catalogue record for this book is available from the British Library.

Library of Congress Cataloging-in-Publication Data
Names: McFadyen, Lesley, editor. | Hicks, Dan, 1972– editor.
Title: Archaeology and photography : time, objectivity and archive /
edited by Lesley McFadyen and Dan Hicks.
Description: London ; New York : Bloomsbury Visual Arts, 2019. |
Includes bibliographical references.
Identifiers: LCCN 2019021064
Classification: LCC TR775 .A715 2016 | DDC 770—dc23
LC record available at https://lccn.loc.gov/2019021064

Typeset by RefineCatch Limited, Bungay, Suffolk

ISBN 13: 978-1-03-223951-4 (pbk)
ISBN 13: 978-1-350-02968-2 (hbk)

For Josh Pollard

CONTENTS

ILLUSTRATIONS

CONTRIBUTORS

Oscar Aldred is a Senior Project Officer at the Cambridge Archaeological Unit.

Joana Alves-Ferreira is CEAACP Junior Researcher in Archaeology at the University of Coimbra.

Iain Anderson is Deputy Head of Architecture and Industrial Survey at Historic Environment Scotland.

J.A. Baird is Professor of Archaeology at Birkbeck, University of London.

Sam Derbyshire is Post-Doctoral Researcher in Archaeology at St John's College, Oxford.

James Dixon is a Historic Environment Consultant for Environment & Infrastructure Solutions UK Ltd, part of Wood. His chapter here was produced when he was Senior Archaeologist and Research Coordinator at Museum of London Archaeology (MOLA).

Sérgio Gomes is CEAACP Junior Researcher in Archaeology at the University of Coimbra.

Alex Hale is Senior Archaeology Research Fellow at Historic Environment Scotland.

Dan Hicks is Professor of Contemporary Archaeology at the University of Oxford, Curator at the Pitt Rivers Museum, and a Fellow of St Cross College, Oxford.

Mark Knight is Senior Project Officer and Director of the Must Farm excavations at the Cambridge Archaeological Unit.

Lesley McFadyen is Senior Lecturer in Archaeology at Birkbeck, University of London.

Antonia Thomas is Lecturer in Archaeology at the University of the Highlands and Islands.

PREFACE

The book has developed from two sessions convened by the editors – first at the Royal Anthropological Institute's conference on Anthropology and Photography at the British Museum in May 2014, and then at the European Association of Archaeologists annual conference in Glasgow in September 2015.

We want to thank Duncan Shields, Melania Savino, Charlotte Young, Colleen Morgan, Danae Fioroe, Ursula Frederick, Carolyn Lefley, Helen Wickstead and Martyn Barber for being part of the conversations that led towards this book, as it moved away from a focus on art towards a more archaeological set of engagements, and to Douglass Bailey for his comments in Glasgow and since. We are indebted to the organizing committees of both of the conferences for facilitating the conversations and exchanges that have led to the current book.

This work would not have been possible without our exposure to the pioneering work in Visual Anthropology and innovative studies of photography in Archaeology and Art History, and we are therefore indebted for inspiration to Elizabeth Edwards, Chris Pinney, Danny Miller, Nick Mirzoeff, Sarah James, Mirjam Brusius, Chris Morton, and Clare Harris. We dedicate this book to Josh Pollard, with thanks for everything over the years.

Lesley McFadyen (London) and Dan Hicks (Oxford)

1

INTRODUCTION: FROM ARCHAEOGRAPHY TO PHOTOLOGY

Lesley McFadyen and Dan Hicks

To press the shutter release is to capture a moment in time. A photograph represents an instant witnessed through the camera lens. It's a fragment of time; a visual remnant; 'a strange stasis, the essence of a stoppage', as Roland Barthes put it (ça-a-été) (Barthes 1980: 142, 176).[1] Before you know it, this freeze-frame begins to behave like an object. It fits into your jacket pocket, your wallet; it slips between the pages of a book. (The archivist even monitors levels of humidity, temperature and light for the stability of these mummified remains, removes steel paper clips and sticky tape.) An optical residue; this collected or preserved experience bursts into the contemporary, flashes up. To point a camera is to make some kind of punctuation. The photograph, as a retrospective image, somehow projects elements that 'rise out from the scene like an arrow and come to pierce' the viewer. It makes 'a little hole, a small clip' (Barthes 1980: 48–49)[2] – more than just a paper-cut from the photocopier. And the analogy follows effortlessly: this optical sherd that digs into the skin of those who look into the past, this beam in the eye, is like an archaeological artefact, a mechanical trace that reproduces the near past, a present absence – and thus a conundrum of simultaneity. Such a 'photographic paradox' must surely (Barthes again) lie in the nature of its content:

> *What is the content of the photographic message? What does the photograph transmit? By definition, the scene itself, literal reality. From the object to its image there is certainly a reduction – in proportion, in perspective and in colour. But at no time is this reduction a transformation (in the mathematical sense of the term).*
>
> *BARTHES 1961: 128[3]*

Archaeology and Photography 'emerged at the same time and under the same interests in the modern world of the first half of the nineteenth century' (Shanks 2016). We have only to recall the earliest photographs, like Roger Fenton's 1857 'Gallery of Antiquities' at the British Museum, for example (Figure 1.1), to see the point. Photography and Archaeology hold in common certain concerns: with objectivity, with a detached and hegemonic modernist viewing of the world, with the visual as evidence, with the stoppage of time for the sake of the monumental. And thus, with some kind of multiplication of time wrought by these 'two devices of western modernity' (Hamilakis, Anagnostopoulos and Ifantidis 2009: 285). Archaeology and Photography: these twin machines of indexicality cast a deictic light across the contours of ruins. The archaeologist's trowel whirrs with the prosthetic gesture of a Polaroid camera, sticking out its flat palm which is a photographic print. The archaeo-photographer holds a chunk of the past in their hand and then, through some trick of reverse chiromancy, reads the vivid and inexorable detail of its unique markings – to predict the past. In these ways and

Figure 1.1 Gallery of Antiquities, British Museum by Roger Fenton, 1857. Albumen print from collodion-on-glass negative, 1857 (Royal Photographic Society, National Media Museum, Science and Society Picture Library).

more, Archaeology and Photography are united in the common endeavour of the representation of history. As if each photograph were a relic, and each pot sherd a snapshot, and each tick of the camera a snap of the clock.

Such – the position in italics above – has been the orthodoxy with which our theme, *Archaeology and Photography*, came to be established as a trope in the archaeo-theory of *fin de siècle* Representational Archaeology. It is an orthodoxy that this book aims to invert. We are thinking beyond the dogma of the contemporary past, the absent present, the sheer artefactualism of a superficial symbolism, bolstered here and there through partial readings of the work of Roland Barthes, which has reduced each side of this pairing to an analogy for the other. Michael Shanks even repurposed Jim Deetz's neologism, *Archaeography*, to express the idea of some post-disciplinary hybridity in the interpretation of the past. Whereas in most areas of cultural studies during the 1980s, Photography came to represent not so much a technical or cultural practice but a theoretical metaphor, in Archaeology it has come to index the methodological fetishization of the artefact as a trace (cf. Krauss 1999: 290; Dubois 2016) – a foundational analogy for the idea that Archaeology represents the past in the present. As if *Archaeology and Photography* were some version of classical reception studies for visual and material culture – as if trapped inside Fenton's *Gallery of Antiquities* forever.

This book experiments with some alternative accounts of observation and participation in Archaeology and Photography. It understands modern technologies and modes of thought as objects of archaeological enquiry, resources for the future not just hegemonies from the past (cf. Hicks and Beaudry 2006; Hicks 2016). It seeks to problematize the artefactualist-presentist assertion of the Representational Archaeology that our discipline works with mere nonhuman ruins and remnants of past moments that are received and interpreted in the symmetry of a contemporary moment (see Shanks 1992; Hamilakis 2008; Bohrer 2011; Hamilakis and Ifantidis 2013, 2016; Carabott *et al.* 2015; cf. Hicks 2003). It understands Archaeology as a visual medium, in which things are discovered and made visible, not just socially constructed or and culturally represented. Rather than reducing Archaeology and Photography to analogies to each other, collapsing one into a metaphor for the other, playing the mirror games of object and subject, of past and present as signifier and signified, we explore some of the different ways in which Archaeology and Photography can live together, and can intervene in the world.

Three main themes are at stake. In each case, we explore how to imagine alternatives to the Representational Archaeology. First, the different approaches to *Objectivity* assembled here start to question Barthes' assertion that the passage from object to image is 'at no time a transformation', imagining the human past and its material evidence as far more contingent (Hicks 2010; Hicks and Beaudry 2010). Second, the question of the *Archive* is framed not in comparative terms, where archaeological remnants are understood to be photographic in quality, but in terms of coexistence and mutual identification of these two methods where archaeological knowledge is unstable (Hicks 2013) – as if, as Bergson put it for science and philosophy, they were 'meant to implement each other' (Bergson 1965 [1922]: 5). And third, a close examination of *Time* in Archaeology and Photography questions ideas of perceptions of short moments of time in the instant and the snapshot, and thus problematizes any notion of constellations of these as 'multi-temporality' (cf. Hicks 2016). If the camera is 'a clock for seeing', as Barthes has it (1980: 33), then we might take the time to listen to 'the living sound of the wood' beyond the instant, in the swing, one tick after another.

*

A generation ago, when the idea of *Archaeology and Photography* emerged in the avant-garde 'post-processual' work, it was a metaphor for the sociology of archaeological knowledge. The metaphor has since developed into a dominant trope in the archaeo-theory of the Representational Archaeology. Photography represented one vehicle, with the veneer of French theory, now Foucauldian, now Barthesian, through which this sociological critique of the archaeological construction of knowledge of the past in the present could be developed:

> Photographs in archaeological texts usually offer either pictorial atmosphere or act as documentary witnesses. The witness says 'I was there'; the photo says 'Look and see'. But looking is not innocent. The eye of the camera, the look with perspective is often the gaze of surveillance, the one-way look. . . It belongs with an attitude which would take the past, appropriate the past, pin it down. Mug shots of the past. Inventories. The atmosphere shot may also speak of the restrained immediacy or spectacle of tourism. The act of looking goes with the meanings it finds. Surveillance finds objects to control.
>
> SHANKS 1992: 145-6

This was an explicitly 'critical' sociocultural approach to the knowledge practices of Archaeology as a discipline:

> Photographs are often taken for granted in archaeology. They are treated as technical aids, helping to record or identify features and objects, or they may provide illustrative ambience, landscape backdrop, evocations of setting. There is little or no questioning of conventional uses of photography. Archaeological photographs are treated as transparent windows to what they are meant to represent. I aim to inspect this apparent clarity. Taking direction

from cultural studies. . . and from the sociology of knowledge. . . my perspective is one of critique, looking within cultural works to reveal sedimented meanings which serve particular interests: it is a negative outlook, aiming, through rational scrutiny, to unveil and debunk neat systems of method and thought, on the grounds that they are always inadequate to reality.

SHANKS 1997: 73

Gradually over the past quarter of a century, this constructivist position has hardened. For example: 'A successful account of the past,' it is suggested, 'is not so much a measure of accordance between the way things were and our archaeological account, as it is a personal and social achievement.' (Shanks 2007: 589-90, 591). The metaphor of the photograph and the artefact as traces of the past has become 'the ontology of the past – that it did happen' (Shanks 2007: 591-2). Archaeology is redefined as 'a field of archaeological work performed upon the material past that persists into the present, involving (media) representations made of the remains of the past' (Shanks and Witmore 2012), reduced to the study of the remains of the past in the present in 'a conjunctive moment of past/present' (Shanks and Svabo 2013: 100). Archaeology is reduced, in other words, to a subfield of classical reception studies dealing with material culture rather than texts (cf. Hicks 2010). The influence of this work has been felt across studies of the archaeological record (Lucas 2012; Carabott *et al.* 2015; Doane 2003: 1; Bohrer 2011: 9).

There are doubtless many problems with how archaeological photography has been practised: the formulaic manner in which photographs are set up and taken on an archaeological site, again and again, serving to flatten out the past through a particular form of visualism (cf. Guha 2002; Lyons *et al.* 2005; Morgan 2016), the monumentalization of sites where the 'clutter' of the archaeological process and the human lives that occupied the space within the frame are cleaned up and erased before the shutter is opened (Hamilakis 2008; Bohrer 2011; Hamilakis and Ifantidis 2013, 2016), the objectification of fieldworkers who are included in the frame (Knight 2000; Baird 2011; Riggs 2016), to name but a few. But attempts to develop alternative photographic practices in archaeology, from 'photo documentaries of theatres of excavation' (Shanks 1992: 184) or 'photo-essays' (Hamilakis and Ifantidis 2013, 2016) to the myriad photographic forms of collaboration between Art and Archaeology, have been of varied success. Such work does not help, and can actively distract from, the problems and the potential of that immense ongoing project of quotidian archaeological practices of creating photographic archives (Baird and McFadyen 2014), those millions of images that fill the museums just as artefacts do, and which archaeologists speak of 'depositing' as if they were not ruins or remnants but new stratigraphic horizons of sediment and fill. When archaeologists speak in these contexts of Photography as documentation, preservation or recording, the Representational Archaeology has taken this at face value, focusing just on interpretation or critique in retrospect as if it were an alternative that could be placed into the slot formerly occupied by explanation. As if the Archaeology were a readymade, or at least already done by others, and our task is only criticism (neither literary, nor Marxist, but simply negative). But this book is filled with archaeologists who do Archaeology (including taking photographs).

Instead, let us recall that the Barthesian punctum was not simply a contemporary moment of interpretation, but 'a roll of the dice' – a form of chance (*hasard*) that can 'bruise and grab' the viewer (Barthes 1980: 49).[4] Our alternative view is to understand Photography not as representation but as transformation, and to find hope in renewing its potential in Archaeology not to perpetuate but to make 'trouble' for civilization:

> To see oneself (other than in a mirror): on the scale of History, this act is recent, the portrait, painted, drawn, or miniaturized, having been, until the diffusion of Photography, a restricted good, destined moreover to display a social and financial standing – and in any case, a painted portrait, however close the resemblance (this is what I am trying to prove) is not a photograph. It is curious that no one has thought of the trouble (for civilization) that this new act causes. I want a History of Looking.
>
> BARTHES 1980: 28[5]

*

On 6 April 1922 at the Société française de philosophie in Paris, Albert Einstein and Henri Bergson debated the implications of relativity for conceptions of time and simultaneity. Einstein argued for a distinction between physical time (measured speed) and psychological time (slower human consciousness of measured time). Bergson's position was that measured speed, no matter how fast its capture, is always derivative in that it requires a spatial translation of time, and that this creates a purely immobile form of movement. 'What is real is the continual change of form', Bergson suggested: 'form is only a snapshot view of transition' (Bergson 1911 [1907]: 319). Time as instant, in this view, is illusory – it is a representation, just as form in photography is represented as static:

> It is true that if we had to do with photographs alone, however much we might look at them, we should never see them animated: with immobility set beside immobility, even endlessly, we could never make movement.
>
> BERGSON 1911 [1907]: 322

Our conception of Archaeology and Photography has remained within an Einsteinian sense of spacetime, imagining that differences in elapsed time are created through the photographic image or the archaeological artefact, in an extreme example of time dilation. The photographic image arrests duration, it is assumed, as a method for standing outside of ongoing, living processes of constant change. For Bergson, this is a fallacy:

> A science that considers, one after the other, undivided periods of duration, sees nothing but phases succeeding phases, forms replacing forms; it is content with a qualitative description of objects, which it likens to organized beings. But when we seek to know what happens within one of these periods, at any moment of time, we are aiming at something entirely different.
>
> BERGSON 1911 [1907]: 351

How can we use Bergson's account of 'the language of transformism' to describe image and object in Archaeology and Photography as more than a still life? How can we invert the idea of unchanging, instantaneous, frozen remnants that multiply time, by understanding these instead as ongoing resources in human life through which time is transformed? Bergson again:

> Continuity of change, preservation of the past in the present, real duration – the living being seems, then, to share these attributes with consciousness. Can we go further and say that life, like conscious activity, is invention, is unceasing creation?
>
> BERGSON 1911 [1907]: 24

It is by recalling the status of Archaeology and Photography as modalities of *knowledge* that we can start to acknowledge the durational, and thus transformational, qualities of images and objects, as distinct from the dogma of representational (instantaneous) accounts that are hooked on the rhetoric of the snapshot and the freeze-frame. Such an approach inverts Foucault's *Archaeology of Knowledge*, which was based on the idea of 'discourse. . .immobilized in fragments, precarious splinters of eternity':

> There is nothing one can do about it: several eternities succeeding one another, a play of fixed images disappearing in turn, do not constitute either movement, time or history.
>
> FOUCAULT 1972 [1971]: 166[6]

Visual Anthropology, and specifically the work of Elizabeth Edwards (see 1999, 2017), is one place where the idea of photography as duration has been developed. Edwards writes about the ways in which the photographic image involves many events and processes (Baird 2011, 2017). This is more than simply asserting that the photograph is some kind of object, or the idea that Visual Anthropology has witnessed some kind of 'material turn' (pace Carabott et al. 2015: 11). It is about photographs as knowledge that endures. Elizabeth Edwards has argued that, 'Photographs here are as much "to think with" as they are empirical, evidential inscriptions' (Edwards 2001: 2). We agree.

Look at the photographs reproduced by the contributors to this book. One is taken in 1930–31 at Dura-Europos, Syria (Figure 4.10). Another taken in 2011 at Porto de Moura 2, Portugal (Figure 6.1). Both document an archaeological excavation, and are stored in the project's archive. They depict landscapes, spoil heaps, people and equipment. In Portugal the archaeologists continue to work whilst the photograph is taken, while in Syria they have stopped work (the people sit and stand) during the time of the photograph. The ongoing movements elide with Sérgio Gomes's account of 'practices through which these redistributions take place' (Chapter 6). He argues that the archaeological photograph holds within it ongoing movements, he asks us to play with the light in the photograph and grasp a new direction in the materialization process. In the relative fixity of the Syrian site, Jen Baird (Chapter 4) notes the shadow over the archaeological remains in the trenches. Precisely because of the photograph, its fixity and shadow, she extends her thinking onto the subject of human relationships on an archaeological site. Who is present as an archaeologist? Who is present in the past?

Archaeological photographs of inscriptions remind us that they are more than just traces. Baird reproduces a series of four photographs of an inscribed altar from Dura-Europos, taken between 1931 and 2011 (Figure 4.8). Compare these with the two photographs taken in 1925 of the Brodgar Stone in Orkney, Scotland discussed by Antonia Thomas (Figures 7.1 and 7.2). Baird notes the shift in concerns, with focus on the inscription over the altar itself. Thomas notes that because the first photograph had the incised decoration chalked in, there is a lasting concern with the incised marks (inscription) over other decoration (pecked, ground and drilled cup marks) or the qualities of the dressed stone itself. For Thomas, the first photograph has set the standard for what is 'actually seen', and she charts the duration and effects of this view. Both authors note the power of the photography of inscriptions in archaeology to construct a singular knowledge of the past, which collapses our readings of text/art and object/architecture. So too with the different forms of modern inscription seen in Alex Hale and Iain Anderson's discussion of modern graffiti at Scalan in Moray (Figures 9.2 and 9.3). The drawing and writing documented in these photographs are more than just traces, and so are the inner connecting wall of the room where sacks of threshed grain are filled, and the space above and on the inner side of the mill door. These images are busy with occupation, and so is the one taken by Philippa Elliott of the room keys on the wall at the reception at Pollphail (Figure 9.5).

Where Hale and Anderson show how photographic images can redefine how we think about the significance of archaeological sites within Scotland's historic environment, James Dixon's discussion of informal photographic archives of a building in the process of transformation – two terraces of early nineteenth-century housing either side of a public house in the London Borough of Southwark (Figures 8.3 and 8.4) – expands our sense of inscription by showing photography's ability to do more than record remnants or ruins in purely nonhuman form. The focus here is human life over time through architectural space. Photographs, artefacts, inscriptions, architecture: all here are ongoing spaces of building inhabitation rather than frozen ruins reconstructed in the present.

There is still more humanity in the photographs of people, and photographs of photographs. Thomas discusses a photograph of the first image of the Brodgar Stone taken in 1925, held in 2011 in the hand of the son of the man who took it (Figure 7.6). Thomas argues that both Archaeology and Photography engage with and create multiple durations, and that this image is a kind of multi-durational object. Meanwhile Sam Derbyshire reproduces his photograph of copies of photographs taken in Turkana in 1930 in the hands of elders from the Nakurio community in 2014 (Figure 10.6). People are sitting, leaning and standing, whilst others are walking into and out of the frame, looking in different directions, looking at photographs, looking at the camera. 'This photograph is not still' – as Hicks argues in relation to his photograph of crayons held by a rubber band at the site of the Calais 'Jungle' in 2016 (Figure 12.5; c.f. Hicks and Mallet 2019). These images are what Mark Knight and Lesley McFadyen in their chapter call 'moving zones', in relation to their photograph of a Bronze Age eel trap in river silts at the site of Must Farm (Figure 3.7). They are not ruins, remnants or traces or frozen moments of time, but animate durations, like any other part of the built environment, full of human life and thought that endures.

*

The chapters in the rest of this volume address the themes of *Archives, Time* and *Objectivity* in a series of different ways. To frame the rest of the chapters, in Chapters 2 and 12 Dan Hicks reassesses the potential of the idea of *Archaeology and Photography*. Taking stock of the idea of the dominance of sociological and 'symmetrical' accounts of photographs as constructions of scientific knowledge, Hicks argues that the idea of 'archaeography' has served to render Archaeology and Photography as metaphors for each other, reasserting the foundational logic of the Representational Archaeology of the 1980s: that archaeology does not discover the past but interprets the past in the present. Approaching the relationship between Archaeology and Photography as a kind of 'identifying with' rather than 'comparing to', Hicks draws upon discussions of photography in the later lectures of Roland Barthes, especially *La préparation du roman* and

Comment vivre ensemble (Barthes 2002, 2003), theoretical approaches from Visual Anthropology and the long history of photographic writing in European cultural studies, and new thinking about socio-digital photography, to introduce the idea of '*Photology*' – the visual knowledge of the past that emerges through Archaeology, and 'in which to see is not to reflect (contemplate, interpret, represent) but to refract (transform)'. Understanding Archaeology as a form of knowledge in which the act of making things visible is central, Dan presents five 'mythologies' in Chapter 12 to explore this idea of photological knowledge, on the themes of *Homonymy, Prefiguration, the Unhistorical, Impermanence, and Appearance*. These mythologies work through a series of close readings of archaeological photographs, including an X-Ray, an artistic work by Indigenous artist Christian Thompson, a rediscovery of a photograph of himself excavating during the early 1990s. Chapter 12 concludes by imagining a 'Transformation of Visual Archaeology' that involves a kind of 'Archaeology as if Photography' – in the place of the contemplative archaeographic interpretation of nonhuman ruins and traces, he imagines Visual Archaeology as a deeply human method for transforming our knowledge of the world.

Mark Knight and Lesley McFadyen address the theme of duration in Chapter 3, from Victorian photography to the photography of archaeological remains at the Must Farm excavations in Cambridgeshire, and to contemporary art. Informed by the work of Henri Bergson, they work through ideas of 'concrete duration', clock time, movement, 'illuminated points' and 'moving zones'. Inverting the idea that Archaeology and Photography share a concern with absence, they explore the ongoing presences in archaeological remains and photographic images. The blur of long exposure times represents human life and movement, as they show in a close reading of 'Bell Street, from High Street' (Figure 3.6a), a carbon print of a calotype photograph by Thomas Annan from the 1860s or 1870s. They call for archaeologists to 'slow down', so that they can attend to duration and the animate in archaeology's images and objects.

Jen Baird (Chapter 4) discusses an archive of excavations at Dura-Europos on the Syrian Euphrates made by Maurice Pillet in the 1920s and now held at Yale University. She uses this archival material to ask, 'What do archaeological photographs reveal about the nature of archaeological time?' She understands the archaeological photograph not just as documentation, but as a transformation of time, creating both time and timelessness. Discussing William Fox Talbot's photograph of the Entrance Gateway to the Queen's College, Oxford in *The Pencil of Nature* (1844), which reveals that it is around 2.40 in the afternoon (Figure 1.2), Baird discusses other kinds of unexpected consequences in photography, such as the inclusion of a shadow that reveals the photographer. We have reproduced one such image as the cover to this book – the shadow of archaeologist Richard Atkinson, and his camera, taking a photo of an archaeological excavation at Dorchester in the 1940s (Figure 1.3). But Baird also

Figure 1.2 Entrance gateway, Queen's College, Oxford, by William Henry Fox Talbot (published as Plate XIII in *The Pencil of Nature*, 1844). As Jen Baird points out in Chapter 4 of this volume, in his discussion Fox Talbot observed how 'sometimes a distant dial-plate is seen, and upon it – unconsciously recorded – the hour of the day at which the view was taken' (1844: 13).

discusses the use of the body for scale and the politics of representation of the bodies of site workers. The humanity of archaeological photographs seen over time is clear: 'a photograph intended in the 1920s to be of an inscription might now be of an archaeological worker', she observes. Photography creates, rather than just documents, archaeological sites, Baird concludes, and the slow and considered tasks of the study of archaeological archives represent a crucial practice for transforming archaeological knowledge, intervening with temporal change, and engaging the politics of the visual representation of the past.

Since 2011, Joana Alves-Ferreira has been taking Polaroid photographs during her excavation work at Castanheiro do Vento, Portugal. In Chapter 5, she explains how this work, experimenting through Archaeology with the idea of the 'instant' photo that brings image and object together in time and space, seeks to transform our conventional sense of objectivity and subjectivity. She introduces the idea of 'parafiction', creating new combinations of the fictional and the documentary – combinations that move beyond the work of the Representational

Figure 1.3 The shadow of archaeologist Richard Atkinson and his camera, taking a photograph of an excavated archaeological feature at Dorchester, Oxfordshire in the 1940s (Ashmolean Museum, AD12946_ATKa_neg612).

Archaeology, which has served to reduce fact to fiction through the idea of constructivism – and documenting the future as much as imagining the past. She concludes that the Polaroid photograph can be understood as a form that thinks – opening up the possibilities of a speculative visual archaeology.

In Chapter 6 Sérgio Gomes explores Archaeology and Photography through the idea of 'poetics'. He observes how photographs from 2011 of an excavation at Porto de Moura 2, a prehistoric site located in the Alentejo in Portugal show nothing of the 'messy and muddy' detail of excavation. Through a discussion of Foucault's account of *The Life of Infamous Men* (1979), he argues that the photographs 'code, frame, translate and communicate experience' and thus can be understood as 'textual spaces' that 'redistribut[e] what is seen and what is unseen'. Gomes uses João Barrento's account of poetry as a powerful tool for making 'notes on what is happening in a given moment at a given place', and through this, understands archaeological photographs, like words, as techniques for making 'sudden memories of the world that create the conditions for change and transformation'. Archaeological photographs are textual, in this view, in that they translate the past.

In Chapter 7, Antonia Thomas examines the idea of 'multiple durations' through the example of the first photograph of the Brodgar Stone, a Neolithic carved stone discovered in Orkney in 1925. Following a discussion of some of the ways in which Archaeology and Photography have been discussed in the past, Thomas describes a sequence of photographic and drawn representations of the Stone, from 1925 to 2005, in each of which slight differences and changes emerge. The act of chalking in the incised marks, made visible for the camera, served to distract from the pecked and ground marks that are also key elements of its prehistoric decoration. This sequence cannot be understood through a conventional archeological account of chronology and 'snapshots' in time, Thomas argues. A discussion of the philosophy of time, from Bergson and McTaggart to Husserl and Alfred Gell leads into an account of how 'archaeology's visual conventions still produce, and perpetuate, a particular conception of temporality'. Both Archaeology and Photography, Thomas argues, 'create multiple durations', and in her account of the contemporary social dimensions of the Stone she shows how these include subjectivity, memory and humanity.

James Dixon, in Chapter 8, considers four different photographic records of a building undergoing redevelopment in Southwark, London. The different photographic archives range from the formal images of a professional archaeological photographer to archaeologists' working shots, images used by the building contractors, and photos taken by artists who used the space for a temporary exhibition while it was empty. Dixon underlines how the archaeological study of buildings has long moved away from an interest only in original architectural intentions, to accommodate more historical and anthropological concerns with the ongoing use and biography of a building over time – the way in which photographs are taken in archaeology has remained focused on ideas of the earliest traces. He shows how a new approach to photographing buildings might accommodate ongoing lived human space – an observation that might also lead us away from seeing photographs as frozen relics of an original past, and towards an interest in their ongoing human durations or 'living record'.

The built environment is also the focus of Chapter 9, in which Alex Hale and Iain Anderson consider the photographic recording of graffiti at two contrasting sites in Scotland – Scalan, which is a nineteenth-century farm and Catholic seminary in the Cairngorms, and Pollphail, which is a 1970s oil-workers village on the west coast of Argyll. The former is the site of a significant body of nineteenth- and twentieth-century informal graffiti made during everyday life at the site, while the latter was the location of a body of graffiti art made by the arts collective Agents of Change, which became a popular destination for 'ruin explorers' between 2009 and 2016. Their interest includes how these visual media relate both to definitions of the significance of these modern sites, and to the methods and practices of photographic recording. They discuss the challenge of recording everyday graffiti on buildings rather than just the original function of buildings, and they identify a commonality between this and the sharing of images of the Pollphail graffiti online, through which public forms of photography can come actively to transform a historic site. Here, a permeability develops between Archaeology and Photography, in which taking a photograph can be compared with inscribing graffiti on a wall, and where archaeological approaches to the built environment can be extended, as suggested by Dixon in Chapter 8, to the visual environment of photography.

Sam Derbyshire expands the book's interest in time in Chapter 10 by considering the problem of ideas of historicity and contemporaneity, on a Western model, in the context of eastern Africa. He brings a distinctively archaeological perspective, developing Elizabeth Edwards's injunction to 'focus on the detail' (Edwards 2001: 2), to the field of 'visual repatriation' and 'photo elicitation' as it has been developed in Visual Anthropology, through an account of research he has conducted in the Turkana region of northern Kenya with historic photographic collections from the Pitt Rivers Museum in Oxford, including photographs taken by Ernest Emley and Wilfred Thesiger. He underlines photography's role, alongside the study of human evolution in the region, in the illusion of Turkana as timeless. Through interviews with Turkana people, he uses the photographs to create new histories of Turkana society, examining not just the images' materiality but what he calls their 'affectivity' – a key element in which the work of visual archaeology represents an intervention in social life.

In the penultimate chapter of the book, Oscar Aldred considers the potential of the archaeological aerial photography – a kind of image in which physical distance, paradoxically, increases what can be seen of the past (Chapter 11). Ranging from the pioneering air photography of O.G.S. Crawford to new digital technologies and the contemporary ubiquity of the aerial image in everyday visual culture, Aldred suggests that the aerial view represents a unique mode of visualism based on the 'accumulation' of actions and times in the landscape, in which the photograph itself operates like another layer of landscape change and time-depth (cf. Hicks and McAtackney 2007). He points out that sites visible as crop marks come and go from visibility according to the season, the weather, the growth of plants, light and the time of day, and so on, so that aerial photographs are repeated again and again by archaeologists, sites alternately coming in and out of vision. Refuting the distinction between modern technology and landscape phenomenology, or between the experience of places on foot or from an aircraft, he draws our attention to the 'ecology' of practices through which these images emerge, and how they transform the world. In Aldred's view, Archaeology is like photography in reverse, in that it takes the time to reveal the layered durations of past times. As visual technologies of the aerial view continually advance, 'we can only imagine', he suggests, 'the transformation in the structures of feeling, and the politics of privacy and surveillance, in how we perceive the world from above that such "archaeological" and aerial developments in visual technology will bring'.

*

Our hope for this volume is that it will contribute to the emergence of a new kind of Visual Archaeology – a body of literature of the same kind of sophistication and cross-disciplinary significance as has developed in the best work in Visual Anthropology (Edwards 2017; Edwards and Morton 2015; Joseph and Mauuarin 2018; Mirzoeff 2011; Pink 2013; Pinney 1992, 2011). In the historical study of archaeological photography, there is already a new generation of exciting and innovative work emerging, for example in the work of scholars like Jen Baird, Mirjam Brusius and Christina Riggs (Baird 2017; Brusius 2016; Riggs 2018). But as long as Archaeology remains held back by the legacy of the use of the archaeo-theoretical idea of Photography as nothing but an analogy, its use as a rationale for the focus of the Representational Archaeology on the trace, the remnant, the nonhuman ruin, received in the present, then this new project cannot begin.

Against the Representational account of Archaeography – where dereliction tourism meets classical reception theory via Latourian constructivism – we offer the notion of the Photological as a way of focusing our attention on the visualism of archaeological knowledge. Where photography was reduced to a metaphor, not just in Archaeology since the 1990s but across cultural studies in the 1980s, we want to reclaim it as a key element of archaeological method and practice. Seven of the eleven chapters in this book are about photographs that were taken by the authors. This is a book filled with practitioners and their collaborators – that kind of Archaeology that you simply cannot do on your own. Just as Fenton's image shows a different kind of Archaeology from that of Atkinson (Figures 1.1 and 1.3), so this book understands Photography not as an analogy but as a tool.

In archaeological practice, in its photographic practice, there is a constant engagement with others' lives and an ability to expand the notion of what a photograph can be (cf. Squiers 2014: 10). Perhaps this is why there are as many (if not more) 'working shots' as formal photographs in archaeological archives? In this same light, although the active creation of alternative photographic practices in archaeology are important, there is also the importance of what is right in front of our eyes (Batchen 2014; cf. Squires 2014), what is already there in archaeological practice and in its photographic archives past and present.

Those three themes again. To understand the past photologically requires us to ditch that depressing nonhuman 'artefactualism' of ruins and remnants and to find new accounts of *Objectivity*, to give up on the rhetorical idea of the past as contemporary and develop more thoughtful accounts of *Time*, and to focus on the camera as intimately related to the *Archive* as a device.

The contributions to this volume are united in an understanding of the archaeological object not as a frozen moment of time but an ongoing human duration, extended through these technologies of the archive and the camera. There is an affinity with Elizabeth Edwards's notion of 'accumulative histories' (Edwards 2001: 13), but a challenge to expand this to accommodate the kind of slow processes of sedimentation that prehistory records, ongoing into the present. Each chapter has extended this observation to the question of photography in a different way: photographs understood as if they were buildings, as if they were people, as if they were memories, as if they were in motion, as if they were a part of oneself, as if they were politics, as if they were community history, as if they were salvage, above all as if they were transformations rather the representations or interpretations. (This is, as Dan's Chapter 12 puts it, Archaeology 'as if' Photography – another reading of Barthes that imagines a speculative alternative to the archaeographic 'ça-a-été'.)

In the nineteenth century, we suggest, Photography as a technology made Archaeology as we understand it today possible. That is the lesson of Fox Talbot's Oxford clock-tower (Figure 1.2). But as Photography itself transforms today, becomes a form of language, a new digital mode of visual knowledge (Hicks Chapter 12 this volume), why should we turn to Archaeology? The socio-digital is surely an unexpected theme for a book about Archaeology.

But Archaeology studies change, and it studies the modern world, and it is a study of what is seen and what is unseen and what might be seen. How a different point of view emerges. It is a means through which we can discover and describe unspoken processes, present prehistories, that are going on around us all the time – unwritten, nonverbal even, and yet far from insignificant. And if 'language is the figure ground reversal of thought' in 'the embodiment of the visual image', as James Weiner has suggested in his discussion of Roy Wagner (Weiner 1995: 18, Wagner 1986), then a visual theory of knowledge might take a form not unlike that that we are sketching for the Photological. And so beyond Archaeology, by showing how the archaeological metaphor of the remnant and the trace can be reoriented today, and how archaeological thinking about material culture is moving on from the old concerns of representation, we certainly hope that the book also makes a contribution to how we understand Photography. Michael Shanks was certainly right to observe that 'archaeology abounds in striking, strange and fascinating images' (Shanks 1992: 1). But our discipline is no antique cabinet of wonder today. It is full of new thinking that can not only change how we think about Archives, about Objectivity, about Time, and thus what *Archaeology and Photography* can mean today – but can also throw light on our socio-digital world of screens and devices, in which the ongoing collapse of event into image, object into subject, of the artificial into the real, of the everyday into the archived, and the interruption of timelines, and the power of making something visible, and the visualism of knowledge, all of this, is strangely familiar to the archaeologist.

With the advent of digital technologies, it is surely improbable, but of course also not impossible, that Archaeology – of all the disciplines! – might be a tool for understanding what the photographic image is today. So let us think ourselves towards an unfolding of visual knowledge that begins with a change of perspective, a transformative Augenblick, a figure ground reversal – *from Archaeography to Photology*.

Notes

1 'une stase étrange, l'essence même d'un arrêt'; 'Le noème de la Photographie est simple, banal; aucune profondeur: *ça-a-été*.

2 'Le second élément vient casser (ou scander) le studium. Cette fois, ce n'est pas moi qui vais le chercher (comme j'investis de ma conscience souveraine le champ du studium), c'est lui qui part de la scène, comme une flèche, et vient me percer. Un mot existe en latin pour désigner cette blessure, cette piqûre, cette marque faite par un instrument pointu; ce mot m'irait d'autant mieux qu'il renvoie aussi à l'idée de ponctuation et que les photos dont je parle sont en effet comme ponctuées, parfois même mouchetées, de ces points sensibles; précisément, ces marques, ces blessures sont des points. Ce second élément qui vient déranger le studium, je l'appellerai donc punctum; car punctum, c'est aussi : piqûre, petit trou, petite tache, petite coupure – et aussi coup de dés. Le punctum d'une photo, c'est ce hasard qui, en elle, me point (mais aussi me meurtrit, me poigne).'

3 'Quel est le contenu du message photographique ? Qu'est-ce que la photographie transmet ? Par définition, la scène elle-même, le réel littéral. De l'objet à son image, il y a certes une réduction: de proportion, de perspective et de couleur. Mais cette réduction n'est à aucun moment une transformation (au sens mathématique du terme)'

4 'Ce second élément qui vient déranger le studium, je l'appellerai donc punctum; car punctum, c'est aussi: piqûre, petit trou, petite tache, petite coupure – et aussi coup de dés. Le punctum d'une photo, c'est ce hasard qui, en elle, me point (mais aussi me meurtrit, me poigne)'.

5 'Se voir soi-même (autrement que dans un miroir): à l'échelle de l'Histoire, cet acte est récent, le portrait, peint, dessiné ou miniaturisé, ayant été jusqu'à la diffusion de la Photographie un bien restreint, destiné d'ailleurs a afficher un standing financier et social – et de toute manière, un portrait peint, si ressemblant soit-il (c'est ce que je cherche à prouver), n'est pas une photographie. Il est curieux qu'on n'ait pas pensé au trouble (de civilisation) que cet acte nouveau apporte. Je voudrais une Histoire des Regards.'

6 Cf. Canales 2009: 181.

References

Baird, J.A. 2011. Photographing Dura-Europos, 1928–1937: An Archaeology of the Archive. *American Journal of Archaeology* 115(3): 427–446.

Baird, J.A. 2017. Framing the Past: Situating the Archaeological in Photographs. *Journal of Latin American Cultural Studies* 26(2): 165–186.

Baird, J.A. and L. McFadyen. 2014. 'Towards an Archaeology of Archaeological Archives.' *Archaeological Review from Cambridge* 29(2): 14–32.

Barthes, R. 1961. Le message photographique. *Communications* 1: 127–138.

Barthes, R. 1980. *La Chambre claire.* Paris, Seuil/Gallimard.

Barthes, R. 2002. *Comment vivre ensemble: simulations romanesques de quelques espaces quotidiens: notes de cours et de séminaires au Collège de France, 1976–1977* (edited by C. Coste). Paris: Editions du Seuil.

Barthes, R. 2003. *La préparation du roman, I et II: cours et séminaires au Collège de France, 1978-1979 et 1979-1980* (edited by N. Léger). Paris: Editions du Seuil.

Batchen, G. 2014. Photography: An Art of the Real. In C. Squiers and G. Batchen (eds) *What is a Photograph?* Prestel: New York, pp. 47–62.

Bergson 1911 [1907]. *Creative Evolution* (trans. A. Mitchell). London: Macmillan.

Bergson, H. 1965 [1922]. *Duration and Simultaneity* (trans. L. Jacobson). New York: Bobbs-Merrill.

Bohrer, F. 2011. *Photography and Archaeology.* London: Reaktion.

Brusius, M. 2016. Photography's Fits and Starts: The Search for Antiquity and its Image in Victorian Britain. *History of Photography* 40(3): 250–266.

Canales, J. 2009. *A Tenth of a Second: A History*. Chicago: Chicago University Press.

Carabott, P., Y. Hamilakis and E. Papargyriou 2015. Capturing the Eternal Light: Photography and Greece, Photography of Greece. In P. Carabott, Y. Hamilakis and E. Papargyriou (eds) *Camera Graeca: Photographs, Narratives, Materialities.* Farnham: Ashgate, pp. 3–21.

Doane, M.A. 2003. *The Emergence of Cinematic Time. Modernity, Contingency, the Archive*. Cambridge, MA: Harvard University Press.

Dubois, P. 2016. Trace-Image to Fiction-Image: The Unfolding of Theories of Photography from the '80s to the Present. *October* 158: 155–166.

Edwards, E. 1999. Photographs as Objects of Memory. In: M. Kwint, C. Breward and L. Aynsley (eds) *Material Memories. Design and Evocation*. Oxford: Berg, pp. 221–236

Edwards, E. 2001. *Raw Histories: Photographs, Anthropology and Museums*. Oxford: Berg.

Edwards, E. 2017. Anthropology and Photography. In T. Sheehan (ed.) *Grove Guide to Photography.* Oxford: Oxford University Press, pp. 127–132.

Edwards, E. and C. Morton 2015. Between Art and Information: Towards a Collecting History of Photographs. In E. Edwards and C. Morton (eds) *Photographs, Museums, Collections.* London: Bloomsbury, pp. 3–23.

Foucault, M. 1972 [1971]. *The Archaeology of Knowledge* (trans. *A.M. Sheridan Smith).* London: Tavistock.

Foucault, M. 1979. The Life of Infamous Men (trans. P. Foss and M. Morris). In M. Morris and P. Patton (eds) *Michel Foucault: Power, Truth, Strategy*. Sydney: Feral Publications, pp. 76–91.

Guha, S. 2002. The Visual in Archaeology: Photographic Representation of Archaeological Practice in British India. *Antiquity* 76, 93–100.

Hamilakis, Y. 2008. Monumentalising Place: Archaeologists, Photographers, and the Athenian Acropolis from the Eighteenth Century to the Present. In P. Rainbird (ed.) *Monuments in the Landscape: Papers in Honour of Andrew Fleming*, Stroud: Tempus, pp. 190–198.

Hamilakis, Y., A. Anagnostopoulos and F. Ifantidis 2009. Postcards from the Edge of Time: Archaeology, Photography, Archaeological Ethnography (a Photo Essay). *Public Archaeology* 8(2–3): 283–309.

Hamilakis, Y. and F. Ifantidis 2013. The Other Acropolises: Multi-temporality and the Persistence of the Past. In P. Graves- Brown, R. Harrison and A. Piccini (eds) *The Oxford Handbook on the Archaeology of the Contemporary World*. Oxford: Oxford University Press, pp. 758–781.

Hamilakis, Y. and F. Ifantidis 2016. *Camera, Calaureia: An Archaeological Photo-ethnography* Oxford: Archaeopress.

Hicks, D. 2003. Archaeology Unfolding: Diversity and the Loss of Isolation. *Oxford Journal of Archaeology* 22(3): 315–329.

Hicks, D. 2010. The Material-Cultural Turn: Event and Effect. In D. Hicks and M.C. Beaudry (eds) *The Oxford Handbook of Material Culture Studies.* Oxford: Oxford University Press, pp. 25–98.

Hicks, D. 2013. Four-field Anthropology: Charter Myths and Time Warps from St Louis to Oxford. *Current Anthropology* 54(6): 753–763.

Hicks, D. 2016. The Temporality of the Landscape Revisited. *Norwegian Archaeological Review* 49(1): 5–22.

Hicks, D. and M.C. Beaudry 2006. Introduction: The Place of Historical Archaeology. In D. Hicks and M.C. Beaudry (eds) *The Cambridge Companion to Historical Archaeology.* Cambridge: Cambridge University Press, pp. 1–9.

Hicks, D. and M.C. Beaudry 2010. Material Culture Studies: A Reactionary View. In D. Hicks and M.C. Beaudry (eds) *The Oxford Handbook of Material Culture Studies,* Oxford: Oxford University Press, pp. 1–24.

Hicks, D. and L. McAtackney 2007. Landscapes as Standpoints. In D. Hicks, L. McAtackney and G. Fairclough (eds) *Envisioning Landscape: Situations and Standpoints in Archaeology and Heritage.* Walnut Creek, CA: Left Coast Press (One World Archaeology), pp. 13–29.

Hicks, D. and S. Mallet 2019. *Lande: the Calais 'Jungle' and Beyond.* Bristol: Bristol University Press.

Joseph, C. and A. Mauuarin 2018. Introduction. L'anthropologie face à ses images. *Gradhiva* 27: 4–29.

Knight, M. 2000. Camera Obscura: Archaeology and Photography. Unpublished paper presented at Garrod Research Seminar, McDonald Institute for Archaeological Research, Cambridge.

Krauss, R. 1999. Reinventing the Medium. *Critical Inquiry* 25(2): 289–305.

Lucas, G. 2012. *Understanding the Archaeological Record.* Cambridge: Cambridge University Press.

Lyons, C., J.K. Papadopoulos, L.S. Stewart and A. Szegedy-Maszak 2005. *Antiquity and Photography: Early Views of Ancient Mediterranean Sites.* Los Angeles: Getty Publications.

Mirzoeff, N. 2011. *The Right to Look: A Counterhistory of Visuality.* Durham, NC: Duke University Press.

Morgan, C. 2016. Analog to Digital: Transitions in Theory and Practice in Archaeological Photography at Çatalhöyük. *Internet Archaeology* 42. http://intarch.ac.uk/journal/issue42/7/index.html

Pink, S. 2013. *Doing Visual Ethnography.* London: Sage.

Pinney, C. 1992. Parallel Histories of Anthropology and Photography. In E. Edwards (ed.) *Anthropology and Photography, 1860–1920.* New Haven, CT: Yale University Press, pp. 74–95.

Pinney, C. 2011. *Photography and Anthropology.* London: Reaktion.

Riggs, C. 2016. Shouldering the Past: Photography, Archaeology, and Collective Effort at the Tomb of Tutankhamun. *History of Science* 55(3): 336–363.

Riggs, C. 2018. *Photographing Tutankhamun: Archaeology, Ancient Egypt and the Archive.* London: Bloomsbury.

Shanks, M. 1992. *Experiencing the Past.* London: Routledge.

Shanks, M. 1997. Photography and Archaeology. In B.L. Molyneaux (ed.) *The Cultural Life of Images: Visual Representation in Archaeology.* London: Routledge, pp. 73–107.

Shanks, M. 2007. Symmetrical Archaeology. *World Archaeology* 39(4): 589–596.

Shanks, M. 2016. Archaeography. Web page from mshanks.com recorded at archive. org on 19 November 2016. https://web.archive.org/web/20161119023436/http:// www.mshanks.com/archaeology/archaeography/

Shanks, M. and C. Svabo 2013. Archaeology and Photography: A Pragmatology. In A. González Ruibal (ed.) *Reclaiming Archaeology: Beyond the Tropes of Modernity*. Abingdon: Routledge, pp. 89-102.

Shanks, M. and C. Witmore 2012. Archaeology 2.0? Review of Archaeology 2.0: New Approaches to Communication and Collaboration [Web Book]. *Internet Archaeology* 32. https://doi.org/10.11141/ia.32.7

Squiers, C. 2014. What is a photograph? In C. Squiers and G. Batchen (eds) *What Is a Photograph?* New York: Prestel, pp. 1–46.

Wagner, R. 1986. *Symbols that Stand for Themselves*. Chicago: Chicago University Press.

Weiner, J.F. 1995. *The Lost Drum: The Myth of Sexuality in Papua New Guinea and Beyond*. Madison: University of Wisconsin Press.

2

THE TRANSFORMATION OF VISUAL ARCHAEOLOGY (PART ONE)

Dan Hicks

Photology, n. *The branch of physics that deals with the properties and phenomena of light; optics.* **photological** *adj.* rare. **photologist** *n. rare an expert in the properties and effects of light.*

Oxford English Dictionary

I. Archaeography

Archaeology And Photography – but on what terms? The conjunction, that 'And', stands as 'a kind of zero-relator, a relational *mana* of sorts – the floating signifier of the class of connectives – whose function is to oppose the absence of relation, but without specifying any relation in particular' (Viveiros de Castro 2003: 1).

Eduardo Viveiros de Castro's comments recall Nietzsche's condemnation, in his *Götzen-Dämmerung*, of the misuse of that conjunction to forge a superficial association:

> There is also a notorious 'and' that I do not like hearing: the Germans say 'Goethe and Schiller' – I am afraid they even say 'Schiller and Goethe' . . . Don't people *know* this Schiller yet? – There are even worse cases of 'and'; I have heard with my own ears, although only from university professors, 'Schopenhauer and Hartmann'.

NIETZSCHE 2005 [1889]: 200[1]

Roland Barthes also recalled these comments by Nietzsche, in his discussion of Proust in the opening lines of his 1978 lecture *Longtemps, je me suis couché de bonne heure*[2]:

Some of you will have recognized the phrase which I gave as the title for this lecture: 'Time and again, I have gone to bed early. Often enough, my candle just out, my eyes would close even before I had time to realize, "I'm falling asleep." And half an hour later, the thought that it was time to go to sleep would wake me': it is the beginning of *In Search of Lost Time*. Does this mean I am offering you a lecture 'on' Proust? Yes and no. It will be, if you like, Proust and Me. How pretentious! Nietzsche spared no irony about the Germans' use of the conjunction 'And': 'Schopenhauer and Hartmann,' he mocked. 'Proust and Me' is even worse. I would like to suggest that, paradoxically, the pretentiousness drops away from the moment when it is me who is speaking, and not some witness: because by setting Proust and myself on the same line, I mean in no way that I compare myself to this great writer: but, in a completely different way, that I identify with him: a confusion of practice, not of value.

BARTHES 1984 [1978]: 313–14[3]

To use Barthes' terms, Archaeology has very often been 'compared to' Photography. Since the 1990s the idea of Archaeology *And* Photography has been a dominant trope of archaeological theory used to suggest what might be shared between the two – in history, in process, and in cultural disposition – in order to make a wider set of arguments about the nature of the archaeological record. For example, Michael Shanks and Connie Svabo describe 'Archaeology and Photography' as possessing a 'common structure' for representing 'old things', invoking James Deetz's (1991) term 'archaeography':

We pursue, even force open the associations between archaeology and photography. The two are usually seen as quite distinct, though related: treated as a technical practice, photography is a long-standing and key component of archaeological documentation. We will offer instead reflection on a hybrid: archaeography. [. . .] We see the connection between photography and archaeology as much more than between technique/medium and discipline. This is a paradigmatic association: archaeology and photography share a common structure, or indeed an ontology. They are homologous to the extent that it is not inappropriate to speak of archaeography. This term also suggests associations beyond archaeology and photography, encompassing both *ta archaia*, old things, and *graphé*, their inscription, record, documentation. Both archaeology and photography are, we propose, aspects of a sensibility, a set of creative practices, certain dispositions towards things [. . .] The focus of this sensibility and constitutive imagination is the persistence of the past, the articulation of remains of the past with the present, recollecting, as a memory practice, bringing what is left of the past before the present, and so involving a dynamics between presence and absence.

SHANKS and SVABO 2013: 89–91

Likewise, Fred Bohrer's art historical account of *Photography and Archaeology* evokes 'the "archaeologizing" effect of photography', understanding Archaeology as 'a paradigm of the challenge of decoding that reigns in photography'.[4] 'There is much to be gained,' chime Yannis Hamilakis, Aris Anagnostopoulos and Fotis Ifantidis, 'by studying the links between photography and archaeology':

> as devices of Western modernity that came into existence more or less at the same time, and partook of the same ontological and epistemological principles . . . If the fundamental event of modernity is the reframing and capturing of the world as picture, as suggested by Heidegger, then both photography and modernist archaeology partook of this process of visualization and exhibition. Both shared the epistemological certainties of Western modernity, be it the principle of visual evidential truth ('seeing is believing'), the desire to narrate things 'as they really were', or objectivism. Both archaeology and photography objectified, in both senses of the word: archaeology produced, through the selective recovery, reconstitution and restoration of the fragmented material traces of the past, objects for primarily visual inspection. Photography materialized and captured a moment, and produced photographic objects to be gazed at. But they both also partook of the modernist inquiry on the *individual and national self as other*, as something external that can be materialized in objects and things, gazed at, dissected and analysed. They also both attempted to freeze time: photography by capturing and freezing the fleeting moment and archaeology by arresting the social life of things, buildings and objects, and attempting to reconstitute them into an idealized, original state.
>
> HAMILAKIS *et al.* 2009: 285

Although it has undoubtedly been suggestive, this hyperconfluential approach has through this Barthesian *comparison-with*, this *confusion of value*, served to abstract Archaeology and Photography as analogues for one another. 'Archaeography' has thus performed that type of mythography that Barthes called *ninisme*: a 'neither-norism' in which 'two opposites that balance each other only in formal terms are relieved of their specific weight' (Barthes 1957: 227).[5] These metaphorical reductions and value confusions have performed a very specific task for that strand of interpretive or post-processual archaeological theory that is best described as the 'Representational Archaeology', since its aim was to 'reconstruct' and 'produce a meaningful representation of the past' (Shanks and Tilley 1987: 20; Shanks 1992).[6] They have reduced Archaeology and Photography to a list of static commonalities: their shared aesthetics, their joint relationships with European modernity, their similar associations with colonialism and ideas of the nation state, a certain ambivalence towards their status as realism or artifice or nature or culture or science or art, but above all the mutual idea of their incessant presentism – a sense of experiencing the past in the present through artefacts or images as remnants that represent fragmented moments of frozen time. This archaeographic sense of the remnant or trace was the key trope of the Representational Archaeology. The foundational metaphor of the archaeographic trace asserts the contemporaneity of the past grounded in a distinctive *artefactualism* that theorizes the image/object as a remnant received from the past in the present, by recalling a familiar 1960s analogy of the photograph to a kind of material trace – one seen for instance in the connections that Barthes made between

> the photograph (the raw witness of 'what was here'), reportage, exhibits of ancient objects (the success of the Tutankhamen show makes this quite clear), the tourism of monuments and historical sites.
>
> BARTHES 1968: 87–8[7]

Thus, for example, Philip Carabott, Yannis Hamilakis and Eleni Papargyriou understand photographs as

> material and mnemonic traces of the things, events, instances and sensorial occasions experienced. They are traces, not in the sense of an imprint, but in the sense of a material remnant, of a relic. It is as if the technological apparatus of photography had managed to extract a material fragment from the 'flesh' of the world, and preserve it for posterity.
>
> CARABOTT *et al.* 2015: 10

This 'artefactualism' with which the Representational Archaeology approached visual and material culture brought two cross-disciplinary elements into play in archaeological theory. On the one hand it made an appeal to certain cultural and semiotic sensibilities associated with what Philippe Dubois (2016) has called the 'effervescence' around the notion of the *photographique* after Roland Barthes' *La Chambre claire*, through which the analogy between artefacts and photographs as traces was drawn (Shanks 1992: 146), just as across disciplines 'photography had left behind its identity as a historical or an aesthetic object to become a theoretical object instead', akin, as Rosalind Krauss has observed, to one of Barthes' *Mythologies* (Krauss 1999: 290). On the other hand, it questioned the objectivity of both image and object through the elements of the 'strong programme' of constructivism in the sociology of scientific knowledge and science and technology studies (Shanks 1992: 80; Shanks 1997: 82).

Thus, Brian Molyneaux's landmark volume *The Cultural Life of Images: Visual Representation in Archaeology* explored how archaeological images – whether a reconstructed scene, a drawn illustration or a photograph – might be expected to stand for vast archaeological periods such as the Upper Palaeolithic through 'the compression of time and space into a single image', and how these representations can stick, being repeatedly reproduced in articles or textbooks. Molyneaux's description of the 'tremendous intertia, or staying power' of the archaeological image (Molyneaux 1997: 6), bore strong similarities to the logic of Bruno Latour's (1986) account of 'immutable mobiles' – those durable forms of the 'inscription' of scientific knowledge that Latour associated with the emergence of modern modes of knowledge. So too Michael Shanks's account of archaeological 'photo-works' as 'the work of making truth' (Shanks 1992: 145) drew directly on this vintage of Latourian constructivism.[8]

Since the turn of the millennium this strand of Representational Archaeology has come to inform further accounts of the role of archaeological material and visual culture in the construction of knowledge of the past more generally. Thus, Gavin Lucas's discussion of archaeological fieldwork extended the account of archaeological 'photoworks' as 'materializing practices' by arguing that 'through the selectivity of the frame and the juxtaposition of fragments, photography makes a new reality, one that does not refer outside of itself' (Lucas 2004: 211; cf. Lucas 2012: 232, 242–3). Jonathan Bateman's pioneering studies of informal excavation photography pushed the idea of the construction of archaeological knowledge and photographic practice further (Bateman 2005, 2006), while Andrew Cochrane and Ian Russell called for 'the development of a critically reflexive practice of visual archaeological expressionism' (Cochrane and Russell 2007: 3), and Carl Knappett extended the use of semiotics in his account of the photograph as 'a composite of icon and index' (Knappett 2002: 106). Meanwhile, the self-styled 'Symmetrical Archaeology' of the late noughties ushered in an updated body of Latourian actor-network theory and STS to expand constructivist claims of 'the disadvantages of photography as technique and the advantages of photography as a practice' through a shift in emphasis 'from visual media as material forms (graphics, maps, photographs) to the work that visual media perform in archaeology' (Witmore 2009: 530; Shanks and Webmoor 2013: 87).

To pick up one thread among the weft of theoretical influences, a handful of early writings by Barthes were central to the development of this view (Shanks and Tilley 1987: 78, 84, 88, 116, 147, 172, 208, 253, 267–8; Barthes 1957, 1964a, 1967, 1970a, 1970b, 1971), and their logic continues to shape this artefactualism today, as the analogy between archaeological remains and snapshots, fragments of time that are 'given form' awaiting reinterpretation, persists:

> *Photography* – turning your relationship with something that interests you, is before you and the camera, into a form that you can take away and look at later, share with friends. *Archaeology* – taking old things that interest you away from where they were found and turning them into an archive, a collection, that you can study, write about, and share. *Archaeography* – giving

visual, (photo)graphic form to such relationships with the remains of the past in a recognition that both archaeology and photography emerged at the same time and under the same interests in the modern world of the first half of the nineteenth century.

SHANKS 2016: n.p.

This representational account of 'the remains of the past' owes much to Barthes' early writing on 'the photographic message', his definition of the photograph as a 'raw witness', his suggestion that 'in the photograph. . .the relationship of signifieds to signifiers is not one of "transformation" but of "recording"' (Barthes 1964b: 46),[9] his idea of a paradox within the photograph that derives from its status as 'a message without a code' (Barthes 1961: 128).[10] In the archaeography of the Representational Archaeology a blank, unstructured 'And' produced an 'excessively assimilative form' (Barthes 2013: 9): an 'artefactualist' model of the presence of the past that centres on the metaphor of the trace, remains, the remnant as a modern concern.

Against this state of affairs, might we re-read Barthes' later account of the conjunction, from that period in which he was shifting his attention 'from texts onto life' (Blonsky 1985: xiv), with which we began? Reimagine Archaeology and Photography as not *comparing-to* but *identifying-with*? As a 'confusion of practice' rather than one of value? What alternative might this offer to the artefactualist account of the material trace? To the claim by the Representational Archaeology to contemporaneity? To the trope of reception? To the reduction of Archaeology and Photography to interpreting and representing momentary fragments from the past in the present?

*

II. Idiorrhythmy

Archaeology and Photography: The Representational Archaeology compared one to the other, and in doing so took both as constants. But Archaeology and Photography coexist in an unstable and long-term relationship rather than representing an eternal state of affairs.

The cracks in the logic of this equilibrial artefactualism run deep, but one place in which they have surfaced in recent years is in the encounter between the idea of analogue relics and the shifting visual landscapes of digital and social media. In their discussion of ideas of 'Archaeology 2.0', Michael Shanks and Chris Witmore dig in, re-stating the representational account of the contemporaneity of the past:

We prefer to think of archaeology less as a discipline and more as an adjective, a field of archaeological work performed upon the material past that persists into the present, involving (media) representations made of the remains of the past.

SHANKS and WITMORE 2012

But any certainty that Archaeology and Photography might *represent* the past through its remains has decomposed as the temporal dimensions of Photography and Archaeology have, since the turn of the millennium, been transformed. These changes in temporal regimes lend an urgency to the task of salvaging for Archaeology and Photography something other than the niche to which the artefactualism of Representational Archaeology has delivered the idea of the archaeographic: ruins, remains, things, traces, 'modern decay', a blank, fragmented and nonhuman materiality (e.g. Olsen *et al.* 2013).

First, Archaeology, through a refreshment and radicalization of its role within the wider project of anthropology, has come to encompass the nonverbal visual and material culture of the most recent past and the near-present (Hicks and Beaudry 2006). Able no longer to define itself according to its object of enquiry ('old things'), any persisting notion of Archaeology as a bilateral encounter with the past 'in or of' the present collapsed more than fifteen years ago (Hicks 2003). (There can be no 'Archaeology *in* and *of* the present'.)

Second, Photography has unfolded through new digital and social technologies, through which an increase in the number of photographs taken each year (from 660 billion in 2013 to 1.2 trillion in 2017) has been accompanied by the gradual loss of any delimited moment or duration of exposure, and thereby an erosion of the distinction between still and moving image forms – in digital technology from the gif to Vine (2012–2016) to an iPhone's 'live photographs', but mainly in social technology, which through new economies of sharing and attention involves a blending of the image with text, comment, metadata, and ongoing processes of modification, resizing, addition, distribution, ephemerality, semi-deletion, crawling, remote harvesting, caching and transactional archiving. As Philippe Dubois has argued, longstanding twentieth-century theories of film – from the Barthesian concept of the punctum as accident (*hazard*) to the Deleuzian distinction between *l'image-mouvement* and *l'image-temps* – have started to give way as the 'temporal elasticity of contemporary images' develops. This is, Dubois argues (2016: 164), 'a change in the regime of photographic visuality' that now annuls any continued assumption of the photograph as an 'image trace':

It is doubtless one of the major characteristics of contemporary modes of the image to change speed ceaselessly, to pass from one regime to the other with suppleness, through continuous variation, without end or any change in nature. Today the fluid scrolling of film no longer radically opposes itself to the freeze frame, as if it were no longer a matter of two contradictory worlds.

DUBOIS 2016: 165

In both cases the effect of these transformations is not just a question of scale, ubiquity or mobility between contexts of creation and consumption. Rather, the status of the objects of Archaeology and Photography as temporally fixed snapshots, remnants, fragments of time is destabilized. In the process, the artefactualist assumption that objects or photographs could be fragments received from the past in the present – that assumption on the part of the Representational Archaeology in which an analogy with a canon of ancient literature theorized by classical reception theory (e.g. Martindale 1993) was always implicit – is compromised.

How then to reimagine this conjunction, that 'And', as neither the sum of two constants nor a timeless reduction to some supposed area of intersection? One concept that is to hand was introduced by Barthes in the first of his lecture series at the Collège de France in Spring 1977. *Idiorrhythmy*, as Barthes has it, envisions two unpredictable rhythms, a coexistence and a relationship that is 'neither dual nor multiple (collective)' where 'cohabitation does not preclude individual freedom' or even isolation, and where a voicing of differences in power and marginality can become possible (Barthes 2013: 6–9, 171, 209). Is it possible, to use the vocabulary that Barthes employed in his discussion of Proust, for Archaeology not to *compare itself to* Photography, but to *identify with* it? This would be to effect that Barthesian 'confusion of practice' by understanding the conjunction as a transformation. To explore Archaeology through Photography so as 'to enter one art through another' (Barthes 2013: 70). To understand Archaeology and Photography not as some subsection of shared endeavour but as 'a play of forces, of mobile intensities' (Barthes 2013: 166). And to begin to perform a reversal of the Representational Archaeology, through a kind of Transformational Archaeology.

Alongside these changes to the temporal regimes of Archaeology and Photography, a third transformation complicates this bilateral space of image and object, performing an unsettling triangulation that holds back any technologically deterministic, socio-digital, presentist explanation of quite how the artefactualist archaeographic model of Representational Archaeology has fallen apart. This transformation relates not to visual and material culture but to thought, for example in the form of a canon of theory. The unstable and unfinished nature of Archaeology and Photography, which are clearly not unchanging categories, ideas, methods, practices or things, is shared by – to run with an example – the writing of Barthes (d. 1980). Unstable and unfinished, that is, not in the sense of a remnant open to endless potential re-readings, reappraisals, reinterpretations, new perspectives, but with precisely the same instability that annuls the artefactualism of any archaeographic theory of the trace. Alongside Archaeology and Photography, then, we might think through Barthes' dialogues with biography, anthropology, visuality, and knowledge, punctuated by his reflections on the Proustian conceptions of time within which photography played such a central role (Shattuck 1963; Brassaï 1997), as a key example of how we might address ideas and knowledge in the same manner as images and objects, for the transformation of Visual Archaeology.

Barthes' best-known works on photography underlined the subjectivities of time and image, personal memory and vernacular cameraworks, in a manner that contrasts with the objectivities that are the province of the pick axe or the camera aperture. But his earlier pronouncements on photography and his book-length account of photography and loss (Barthes 1980) are joined by the less widely discussed thinking introduced in his lectures at the Collège de France between 1977 and 1980 (Barthes 1978, 2002a, 2002b, 2003)[11] written before uses of Barthesian thinking in anglographic archaeotheory and revealed as his posthumous corpus. Through which Barthes has continued to change. With ongoing publication and translation between languages, from notebooks and sound recordings, Barthes' oeuvre has expanded since his death in 1980, in a sequence that has continually jumbled and run out of step with the dates of the writing, most recently at such a scale that Dora Zhang has drawn a comparison with Tupac Shakur, asking, 'Can a hologram Barthes be far behind?' (Zhang 2012).

'What does the photograph transmit?' asked Barthes in 1961:

> By definition, the scene itself, the literal reality. From the object to its image there is certainly a reduction – in proportion, in perspective, and in colour. But this reduction is at no point a *transformation* (in the mathematical sense of the term).
>
> BARTHES 1961: 128[12]

But in retrospect and in the long run, that relationship of transmission was surely always a transformation: preimage to image, latency to bleaching, 'processing', translations and inversions, flips and expansions, rinsing and fixing, printings and distributions. Not just the archaeologist's artefact and the photographer's negative, then, but even the oeuvre of Roland Barthes does not represent a 'raw witness' to be directly received in the present. Just as Barthes suggests that 'the Proustian landscape changes over time', posthumously unfinished and continually shifting shape (Barthes 2011: 209). We find ourselves a long way from any assumption that Barthes might yet retain, or might ever have shared with Walter Benjamin, some view of 'the photograph as a moment of history', which 'mediates a world gone by, graspable at first glance as if through a window directly on to the past' (Dant and Gilloch 2002: 7). A long way from John Berger's account (1966: 583) of photographs as 'relics of the past, traces of what has happened'.

*

Let us suggest that as modes of knowledge the archaeological object and the photograph and also theoretical writing are each subject to a transformative set of contingent deferrals; each leg of the stool 'has been absolutely, irrefutably present, and yet already deferred' (Barthes 1980: 121).[13] Let us hold on to Marilyn Strathern's anthropological account of the 'doorstep hesitation' in cross-disciplinary encounters (Strathern 1987: 286), the prospect of a suspended 'deep hesitation' that 'enables the anthropologist to *not* make connections (start comparing) before the moment is right' (Holbraad and Pedersen 2010: 17). This move from a Representational to a Transformative Archaeology requires a form not of comparison but of 'identification with'. Let us imagine the archive as not a mimetic aftereffect but an anticipated postmodifier, for which 'significance is acquired through the subsequent writing' (Strathern 1999: 9). Archaeology, Photography, Knowledge, each a 'second coming of the world' (Silverman 2015: 33) in which that world is seen to be transformed.

*

III. Histories of looking

Archaeology and Photography have thus been presented as in a relatively timeless and unchanging equilibrium, and this has formed a central strategy in an artefactualist vision of the Representational Archaeology, in which the archaeographic remnant is taken to represent the past in the present. But in other work a renewed historical sensibility has examined the emergence of distinctive regimes of visualism that characterize Archaeology and Visual Anthropology. Might this body of literature begin to address Barthes' desire for 'a history of looking' (Barthes 1980: 28)?[14]

In Archaeology, central here is the work of Stephanie Moser, who has built on the pioneering work of Stuart Piggott (1978) on the 'visualization' of archaeology, a field of enquiry where approaches from the history of art can be brought into a new dialogue with the history of science, through studies of antiquarian illustrations and historic museum displays as forms of 'visual language' and visual conventions for constructing knowledge of the past (e.g. Moser 1992, 2001, 2014; Moser and Smiles 2005). The rich potential of such an approach has been pushed further through Sara Perry's studies of the history of the Institute of Archaeology in London through the theme of visualization (Perry 2011, 2013), Roger Balm's historical sense of 'archaeology's visual culture' (Balm 2016), Jennifer Baird's extension of the idea of 'visual language' to the history of archaeological photography (Baird 2017), Christina Riggs's accounts of the photographing of Indigenous labourers (Riggs 2017a) and the role of photography in the circulation of archaeological knowledge (Riggs 2017b), and Mirjam

Brusius's accounts of archaeological 'decipherment' and objecthood through photography (Brusius 2013, 2016).

In Visual Anthropology, a new generation of historical studies of photographic practice developed from Chris Pinney's account of 'The Parallel Histories of Anthropology and Photography', where the idea that both fields 'appear to derive their representational power through nearly identical semiotic procedures' began on a similar path to that of the archaeographic literature, but has in contrast catalysed a 'creative convergence' between these fields through historical studies (Pinney 1992: 74, 91). Visual anthropology has gradually expanded its scope from ethnographic film-making as a 'visual methodology' (Rose 2000, Banks 2001, Pink 2012) to encompass the study of still photographs, especially through the work of Elizabeth Edwards who has underlined how deep connections with the photographic run into the history of Anthropology as a discipline (Edwards 1992). 'The science of anthropology,' E.B. Tylor observed in his discussion of Carl Dammann's *Anthropologisch-Ethnologisches Album* (1876: 185), 'owes not a little to the art of photography' – and the approaches of Pinney and Edwards have certainly chimed with this nineteenth-century idea of what Anthropology owes to Photography as a mode of objectivity made through observation. Twentieth-century anthropology expanded and accelerated the servicing of this debt beyond recognition, as its focus shifted from the museum archive to the immersions of the ethnographic field. Thus, Malinowski's iconic account of his fieldwork in the Trobriand Islands, in which the use of photography played such a central role (Pinney 2011), sought to allow its reader to 'visualise concretely' (Malinowski 1922: 175) the circumstances of *kula* exchange:

> Right through this account it has been our constant endeavour to realise the vision of the world, as it is reflected in the minds of the natives. The frequent references to the scenery have not been given only to enliven the narrative, or even to enable the reader to visualise the setting of the native customs. I have attempted to show how the scene of his actions appears actually to the native, to describe his impressions and feelings with regard to it, as I was able to, read them in his folk-lore, in his conversations at home, and in his behaviour when passing through this scenery itself.
>
> MALINOWSKI 1922: 229

'You can never take too many photographs,' observed Marcel Mauss (1947: 15).[15] In one famous example of putting this maxim into practice, Gregory Bateson and Margaret Mead made what John Collier (1967: 6) described as the 'saturated' use of photography as a technology; they sought (Edwards 2011: 188) to reveal the otherwise invisible in *Balinese Culture* (1942), shooting 'over twenty-five thousand photographs and twenty-two thousand feet of film'. Since then Anthropology has increasingly come to foreground 'photographs as objects', defining its relationship with material culture studies (Edwards and Hart 2004), but also building on Mead's vision of Visual Anthropology as moving beyond the textual to reveal their status as forms of knowledge wrought through a particular understanding of the 'objective' (cf. Daston and Galison 2007). As Mead put it,

> Relying on words (the words of informants whose gestures we had no means of preserving, words of ethnographers who had no war dances to photograph), anthropology had become a science of words.
>
> MEAD 1975: 5

Anthropological archives have been reimagined as spaces that blend the intellectual histories of comparison and classification with the practical histories of photograph collecting and disposal (Morton 2012; Edwards and Morton 2015; Knowles 2014). Defining the field as the study of *Raw Histories* (Edwards 2001), and of *Photography's Other Histories* (Pinney and Peterson 2003), a new wave of what we might call 'Visual-Historical Anthropology' extended the re-emergence of historical approaches in 1990s anthropology into the study of the visual, drawing freely on this literature's tropes of biography and entanglement. For Elizabeth Edwards, photographs represent 'very literally raw histories in both senses of the word – the unprocessed and the painful' (Edwards 2001: 5). The temporal frame of reference has also spread from the social historical past into the anthropological present, exploring moments where photographs disrupt or 'extend' time (Edwards 2005: 31), collapsing the past into the present to reveal their 'double temporality' (Pinney 2011: 85). The analysis of such archival effects has built on the longstanding anthropological sense of photographs as 'images with a memory' (Collier 1967: 3) – 'a mirror with a memory', as Oliver Wendell Holmes described the daguerreotype in 1859 – and has led to new forms of fieldwork and interpretation, some of which is described as 'photo-elicitation' (Harper 2002) and involves returning copies of historical ethnological photographs to communities, and re-readings 'against the grain' (Coombes 2016). From such work, a new set of questions about how the colonial and postcolonial histories of anthropological photography, as a visual history of unseen lives, can be understood in contemporary political perspective has begun to emerge (Buckley 2014; Lydon 2016).

This turn to history in the study of Archaeology, Anthropology and Photography has much in common with Lorraine Daston and Peter Galison's influential *Begriffsgeschichte* of ideas of *Objectivity*. A special role is assigned to the camera in Daston and Galison's account of nineteenth-century 'mechanical objectivity', in that it comes to stand for a wider shift across a range of scientific media towards resisting the abstractions of ideal types:

> Photographic depiction entered the fray along with X-rays, lithographs, photoengravings, camera obscura drawings, and ground glass tracings as

attempts – never wholly successful – to extirpate human intervention between object and representation.

DASTON and GALISON 1992: 98

This use of a historical perspective to relativize the idea of visual 'objectivity' is undoubtedly an important outcome of weaving the History of Science together with territory more usually explored by the History of Art. But its limits in Visual Anthropology are reached when objectivity is replaced by the same kind of artefactualism as witnessed in the Representational Archaeology. The assertion that we should understand 'photographs as objects' and acknowledge 'the ubiquitous materiality of photographs' has presented the realism of 'photo-objects' as evidence of the subjectivity of the image:

> A photograph is a three-dimensional object, not only a two-dimensional image. As such photographs exist materially in the world, as chemical deposits on paper, as images mounted on a multitude of different sized, shaped, coloured and decorated cards, as subject to additions to their surface or as drawing their meanings from presentational forms such as frames and albums. Photographs are both images *and* physical objects that exist in time and space and thus in social and cultural experience. . . An approach that acknowledges the centrality of materiality allows one to look at and use images as socially salient objects.
>
> EDWARDS and HART 2004: 1, 15

In the face of this invoking of a particular vision of material culture studies in order to turn subjective representation into a sense of the relativity of photographic knowledge, that Barthesian prospect of a history of looking remains undeveloped. Undeveloped, that is, when compared for example with the scope of questions raised by Siegfried Kracauer's philosophical tracing of 'the many existing parallels between history and the photographic media': asking how, in the face of the Rankean ideal of historical reality, *wie es eigentlich war*, a philosophy of history might compare 'historical reality and camera reality' (Kracauer 1969: 3–4). And further, asking how directly the very invention of Photography itself might have influenced the assumption in Diltheyite historicism that it is possible to 'explain any phenomenon purely in terms of its genesis':

> That is, they believe at the very least that they can grasp historical reality by reconstructing the series of events in their temporal succession without any gaps. Photography presents a spatial continuum; historicism seeks to provide the temporal continuum . . . Historicism is concerned with the photography of time.
>
> KRACAUER 1993: 424

'Underlying each new technology is a new optics,' suggested Walter Benjamin (1972 [1936]: 499).[16] Could it be that Photography underwrote the conditions for the emergence of Archaeology? And what then of the development of 'Visual Culture' as not a new subfield within Art History but its fundamental reorientation, effecting a substitution of the cultural for the historical and the visual for the artistic (Foster 1996: 104)? If Photography was complicit in seeing historically, then today we must require a relatively unhistorical approach to Archaeology's changing visualism over time.

*

IV. Visualism

The prospect of a transformation of Visual Archaeology, with which this chapter is concerned, does not require the spectacle of constructing a new auxiliary episteme based on the arbitrary selection of one human qualifier – another archaeo-adjective to join Prehistoric, or Mayan, or Maritime, or Public, etc. Nor will it involve exhausting at some future point the full range of sensory archaeologies, from the Tactile to the Olfactory.[17] What it does require is a turning away from visualization – 'the formation, in the mind, of images or representations of the world' (Ingold 2000: 282) – towards other ways of seeing as archaeologists; giving up on any fantasy of some postdisciplinary *Bildarchaeologie* of the interpretation of a ruined past. In the place of a representational emphasis on visualism as a key element of modernity shared between Archaeology and Photography, this chapter imagines a transformational and altogether less nonhuman sense of how, just as Archaeology and Photography have modern histories, so visualism is not stable in form: over time, from place to place, or across disciplines. Nick Mirzoeff, for example, identifies three shifting 'complexes of visuality' – from plantation slavery, to imperialism, to the present-day military-industrial complex (Mirzoeff 2011) – while many others have traced some of the roles that can be played by visual culture in knowledge (Griffiths 1997) and power (Mukerji 1984).

We might reimagine Archaeology's regime of visualism, I suggest, as *a complex of transformation*. In doing so we might shift our concern from the interpretation or representation of the past to documentation as intervention, by which I mean a concern not so much with what has fallen out of sight as with what can be made visible, whether buried or unburied. The argument is perhaps best expressed by reclaiming Jim Deetz's neologism 'archaeography' from its appropriation by the Representational Archaeology, recalling how he originally defined it:

> By archaeography I mean the writing of cultural descriptions based on the material record, buried or not, in a fashion analogous to ethnographic

description. Archaeography stands in the same relationship to ethnology, the study of culture holistically, as does ethnography . . . And while I'm not terribly taken with the term, it does have its uses.

DEETZ 1991: 430

Following Deetz, let us define archaeological objects, image and knowledge as emerging through the performance of research undertaken with the archaeologist's body in a refraction of the culturally visible that transforms the unseen into the observed. Archaeology's visualism 'transcends human vision' in that it operates across time as Kracauer describes camerawork operating across the 'fortuitous combinations which represent fragments rather than wholes' (Kracauer 2012 [1951]: 187, 191). The visual knowledge that results does not take the forms of scientific detachment or distanced contemplation usually emphasized by historians of science; it instead changes what can be seen, through excavation or the many other forms of discovery that constitute archaeology's distinctively transformative mode of observation.

The Representational Archaeology always insisted, now less of a mantra than a cliché, that our discipline should have long ago abandoned outdated concerns with the prospect of discovering, uncovering or revealing the past. Half a dozen examples from among so many will suffice:

- 'Archaeologists do not discover the past but take shattered remains and make something of them,' was Shanks's constructivist assertion (Shanks 1996: 4).

- Again: 'Archaeologists do not happen upon or discover the past. Archaeology is a process in which archaeologists, like many others, take up and make something of what is left of the past. Archaeology may be seen as a mode of cultural production' (Shanks and Pearson 2001: 50).

- 'Archaeologists . . . do not discover the past. Archaeologists work on what is left of the past,' Shanks repeated (2007: 591). 'Again, archaeologists do not discover the past, but treat the remains as a resource in their own creative (re)production or representation,' he underlined (2007: 592).

- 'Archaeologists do not discover the past,' echoed Tim Webmoor (2013: 109).

- Another minor variant: 'archaeology does not discover the past *as it* was but must work with what remains; archaeologists work with what has become of what was; what was, as it is, always becoming' (Olsen *et al.* 2013: 6)

- Yet again: 'Archaeologists do not discover the past as it was; they work on what becomes of what was, and they work with old things in order to achieve particular ends' (Rathje et al. 2013: 5).

But the claim of the Representational Archaeology that it could work on 'old things' and remnants without discovery, mistakes the archaeological archive for some disembodied readymade (Hicks 2016a: 37). (Something surely more nuanced than such armchair abstraction was contained in Barthes' account of how 'Our gaze can turn, not without perversity, to certain old and beautiful things which now signify something abstract, out of date'. [Barthes 1978][18].) But how to think through a type of archaeological visualism that is more than mere contemplation?

Julian Thomas's neo-Foucauldian analysis of Archaeology and Modernity repeated the representational preference for surface above depth, witnessing without discovery, but it did in the process recall a crucial body of debate about visualism that emerged in landscape archaeology in the early 1990s, during the period in which new approaches in cultural geography were influencing archaeological theory (Thomas 2003: 149–50).[19] Here, critiques of 'the politics of vision' extended discussions of the 'authoritarian gaze' in feminist geography (Massey 1991), which not only drew from Foucault but also from pioneering feminist film studies of the directionality of the 'male gaze' and the use of 'subjective camera' (Mulvey 1975: 11, 16), observing how distanced forms of visuality were bound up with colonial legacies, inherently masculine, and serving unhelpfully to separate off nature from culture, performing what Donna Haraway called the 'god trick' of appearing, quite impossibly of course, to view the world from nowhere (Bender 1999; Hicks and McAtackney 2007; Haraway 1991: 189). The resulting critiques combined gendered, ethnographic and phenomenological accounts of the archaeological gaze as one-way and ego-centred, as Western, as just one (detached) possible form of the perception of the past (Bender 1993; Thomas 1993). These critiques provided a helpful counterpoint to the most auteurist instincts of the Representational Archaeology towards diminishing the discipline, most notoriously in that twist of the archaeological imagination that drove Michael Shanks's seven-paragraph account of a visit to a strip-club in which the author called for an 'archaeological erotics' using an extended comparison between excavation and déshabillement in which 'discovery is a little release of gratification' (Shanks 1992: 54–5). The 'insidious language and images' (Gilchrist 1992: 191) and voyeuristic Schaulust that ran through that bleak parody of Roland Barthes' analysis of the Moulin Rouge in Mythologies (1957: 137–40) is a small reminder of how easily the reduction of archaeological visualism to representation gives way to objectification.

Similar concerns ran through anthropological debate about disciplinary visualism too, among which Johannes Fabian's classic critique of the politics of 'ocularcentrism' in social anthropology forms an interesting parallel. Situating it in a longer-term context of scientific thought and practice since the early modern period, Fabian criticized anthropology's 'visualism' as an epistemological technique for othering, a form of representation allied to the denial of the coevalness of others:

> a cultural, ideological bias toward vision as the 'noblest sense' and towards geometry qua graphic-spatial conceptualization as the most 'exact' way of communicating knowledge. . ... Visualism may be a symptom of the denaturation of visual experience.
>
> FABIAN 1983: 106

Despite this critique, and quite apart from the work of visual anthropology itself, there are a number of approaches in anthropology that, unlike Fabian, place significant value on anthropology's visualism and the discipline's model of 'observation' as something other than a metaphor for witnessing or describing for the purpose of representation.

More recent advocates of the primacy of visualism in Anthropology have included Tim Ingold who, through the primary influence on his work of psychologist James Gibson's account of 'ecological optics' in his book *The Ecological Approach to Visual Perception* (Gibson 1979), has grounded his anthropological account of the environment in the theme of perception (Ingold 2000). In the ongoing influence of the notion of 'affordances', it has sometimes been forgotten that Gibson coined this term for a theory of vision, to answer the question: 'If there is information in light for the perception of surfaces, is there information for the perception of what they afford?' (Gibson 1979: 127), before theorizing 'three main types of going out of sight': occlusion, distance, or darkness (Gibson 1979: 308). The prospect of an archaeological sense of visualism that can work with the unseen as well as the seen holds much in common with how Rane Willerslev (2007) has taken stock of Fabian's critique of anthropology's visualism by taking forward Ingold's vision of a sensory anthropology in which there is more than object and surface, and where

> the primacy of vision cannot be held to account for the objectification of the world. Rather the reverse; it is through its co-option in the service of a particular modern project of objectification that vision has been reduced to a faculty of pure, disinterested reflection.
>
> INGOLD 2000: 235

Ingold proceeded to make a new case for visualism, understanding sight as a bodily practice like any other:

The case against vision is comprehensively disproven. Indeed, it should never have been brought in the first place. It is as unreasonable to blame vision for the ills of modernity as it is to blame the actor for crimes committed, on stage, by the character whose part he has the misfortune to be playing.

INGOLD 2000: 235, 287

Drawing on Merleau-Ponty's account of *The Visible and the Invisible*, Willerslev explores the implications for anthropology of the idea that if 'to see is to have at a distance' then distance 'is not an obstacle to seeing but a precondition' (Willerslev 2007: 26). In keeping with an emerging recognition that anthropology's forms of detachment can take different forms, some of which are not necessarily negative (Candea *et al.* 2015), Willerslev reimagines how the process through which 'vision sets and maintains a distance' can represent 'a sort of depth reflexivity (a seeing that not only sees the world, but turns back upon itself, making itself visible)', a 'mirror-sense of vision . . . as a kind of defence mechanism against the dissolution of the self' in field research (Willerslev 2007: 41). Willerslev thus problematizes the conventional distinction between scientific and objective detachment on the one hand, and on the other, interpretive critiques of such detachment that only intensify the representational impulse by substituting one kind of distance with another, whether a phenomenological attempt to stand inside the moment to critique the external view, or an attempt to stand outside discipline or fieldwork, for example in the ambiguity of the location from which Shanks and Tilley's complained that 'archaeological history stands before the visitor as fetishized objectivity, a detached objectivity mysterious to the visitor, truly fetishistic' (Shanks and Tilley 1987: 70).

The analytics of disconnection suggested by the anthropological accounts of visualism put forward by Ingold and Willerslev holds much in common with how Archaeology might identify with Photography. The collapse of immediacy and distance that the archaeological image brings is captured by W.G. Sebald's commentary on Thomas Browne's *Hydriotaphia* (1658), which hints that the antiquary's melancholic account of the excavation of particles of cremated human bone in earthenware vessels buried shallow in the Norfolk soil might reveal that in the practice of archaeology 'the more the distance grows, the clearer the view becomes':

You glimpse the tiniest details with the utmost clarity. It is as if you were looking through a reversed telescope and through a microscope at the same time.

SEBALD 1995: 29–30[20]

(Or, as Walter Benjamin has it, 'in the optics of history – in this respect the counterimage of the spatial – movement into the distance means growing larger' (Benjamin 1972 [1927]: 348).[21])

As with Willerslev's account of 'depth reflexivity', this Sebaldian sense of intimate distance, of a collapsing of our conventional scales of both physical and temporal proxemics, may spur us to think further about forms of detachment based on transformative *idiorrhythmy* rather than distanced representation. In doing so we may come, in our description of Archaeology's visualism, closer to Barthes' weaving together of biography and image, of 'the aporia of bringing distances together' (Barthes 2013: 6), as in those *moments de vérité* through which fiction and reality come together, for example in the recognition of one's own circumstances in a novel, in our recognition of Proust in Barthes' discussion of Photography, or in the place of the death of his mother in Barthes' theory of photography:

> It was History that separated me from them. History – is that not simply that time in which we weren't born? I was reading my nonexistence in the clothes my mother had worn before I could remember her.
>
> BARTHES 1980: 64[22]

(Barthes imagines his relationship with Proust by identifying with his response to the 1905 death of Jeanne Clémence Weil (Proust's mother), suggesting that Proust was, like him, 'in search of a form which recollects suffering and transcends it' (Barthes 1984 [1978]: 315).[23])

What form of archaeology's visualism is present here? It is not just an experience of loss but the knowledge of loss. Knowledge that operates like the Barthesian sense of 'recognition' – reading, writing, and 'holding forth'. Knowledge that transcends straightforward questions of authorship or reception by dwelling on transformation across time. If Archaeology's visualism is not simply critiqued, or resisted, or erased, or reduced just to what it shares with Photography, then could it, as Willerslev suggests for anthropology's modes of seeing, be re-envisioned, away from the hostility towards the visual that has so often characterized cultural theory (Jay 1993), as a modality of disciplinary knowledge?

What then is that modality of disciplinary knowledge? If archaeology is a bringing into the field of vision that forms a counterpoint to Gibson's account of occlusion, distance and darkness, then we might term the visual form of knowledge of the past that emerges through its performance of transformation – in which to see is not to reflect (contemplate, interpret, represent) but to refract (transform) – *photology*.

*

V. Photology

Reimagining Archaeology and Photography as the idiorrhythmic coexistence of two techniques for the transformation of the past, rather than some static state of affairs, serves then to problematize any conception of a second bilateral relationship: the ambiguity that is envisioned between the image and what it depicts, the idea that a photograph somehow 'simultaneously presences and surrogates for what is not present' (Ginzberg 2001: xiv), and the by now familiar suggestion that the photograph constitutes 'the paradox of an event that hangs on the wall' (de Duve 1978: 113). The Representational Archaeology of course used a photographic sense of presence and absence as one of their principal metaphors. The mirror games of this artefactualism, this metaphysics of ruin, began with this bilateral framing of time by turning an analogy into a one-way street:

> Certainly the image is not the reality but at least it is its perfect analogue and it is precisely this analogical perfection which, to common sense, defines photography.
>
> BARTHES 1961: 128[24]

In practice, one narrow trope of archaeology's visualism has been reclaimed (cf. González-Ruibal 2013), reviving that base form of archaeo-voyeurism detachment that coloured interpretive archaeology's use of the photographique in the resurrection of the early modern theme of ruins in the 'dereliction tourism' (Mah 2014) and 'ruin porn' (Mullins 2014: 48) of archaeological photography that attends to abandonment and decay and yet in the process wildly dehumanizes change and loss, that uses one peculiar kind of visualism to reduce archaeology to merely 'the science of ruins and the abandoned, of fragments and death' (González-Ruibal 2008: 248). Thus the Representational Archaeology resorted to the insouciant prospect of 'an archaeology of ruins' (Pétursdóttir and Olsen 2014) that was characterized by a sheer metaphorical ocularity (Rorty 1979: 11), in the face of duration a mangling of the Benjaminian account of the transmission of aura. The confidence that a fragment must be a nonhuman trace betrays, to borrow a distinction from Marcel Duchamp, a purely 'retinal' understanding of the work of Archaeology – as opposed to a conceptual one. Where some work in Visual Anthropology collapsed of the idea of a constructed reality represented by the image into the flat assertion of the concrete materiality of the photograph as *tangibilia* (cf. Barthes 2013: 56), so the Representational Archaeology has taken at face value Barthes' assertion that 'in Photography, the presence of the thing (at a certain past moment) is never metaphoric', (Barthes 1980: 123).[25] Consider in this regard Fred Bohrer's account of how 'archaeology functions as a kind of photography', and in particular his claim that

> the sheer material reality of the object in archaeology is the perfect theatre for enactment of what, in Barthes's terms, is photography's 'reality effect'.
>
> BOHRER 2005: 188–9

Bohrer breaks down the history of archaeological photography into two parts based on the status of 'science or truth': 'an early phase of elation at this new vehicle for objective representation followed by a later one of resignation, using it despite its faults' (Bohrer 2011: 29). There is a strangely contracted lurch from a caricature of Victorian naïve positivism – 'the initial expectation that photography would simply and quickly capture distant reality' (Bohrer 2011: 66, 70) – to the promise of a postmodern corrective. ('It was inevitable that someone would attempt to create a broad interface between postmodern ideology and archaeological praxis,' lamented Mike Schiffer [1994: 159]). In a similar vein, Gavin Lucas has distinguished ideas of 'survivals' and 'relics' in Victorian anthropology from the sense of contemporaneity of the material past among the Representational Archaeologists:

> In the 19th century, the archaeological record itself was the anachronism. Today, it is our own subjectivity, imposed on interpretations of the past, that is anachronistic, which in the context of archaeology is principally expressed though the role of analogy.
>
> LUCAS 2015: 7

This notion of a contemporary past of ruins and remnants maps onto Archaeology a distinction also found in Barthes' early accounts of photography and time: between *l'être-là* ('being-there') and *l'avoir-été-là* ('having-been-there'):

> Photography establishes in effect not a consciousness of the being-there [*l'être-là*] of the thing (which any copy could provoke) but a consciousness of its having-been-there [*l'avoir-été-là*]. It is about a new category of space-time: local-immediate [*locale immediate*] and temporal anteriority [temporelle antérieure]; in photography there is an illogical conjunction between the here [*l'ici*] and the then [*l'autrefois*].
>
> BARTHES 1964b: 47[26]

If the 'confusion of value' that compares Archaeology and Photography leads to such artefactualism, what does its reversal – a 'confusion of practice' that 'identifies' one with the other – bring about? Neither substitution nor mimesis nor good red herring, in the alternative approach to Archaeology and Photography suggested here, an archaeological artefact, a photographic image, or a theoretical text are ongoing and open-ended transformations – of here and then and now and there. In this view, a concern with optics is a concern with concepts, with human knowledge rather than material ruins.

Against Lucas's association of relics and survivals with early ethnology, let us recall how the technical and dioptrical qualities of light were among the early intellectual concerns of the two key founders of British and American anthropology.

E.B. Tylor's paper on 'Refraction of Light Mechanically Illustrated' was published in 1874 in *Nature* and repeated John Herschel's comparison of light rays with how 'a line of soldiers marching across a tract of land' could be 'defracted' by rough ground (Tylor 1874: 158–9).[27] Seven years later at the University of Kiel, Franz Boas's doctoral thesis, *Beiträge zur Erkenntnis der Farbe des Wassers* addressed questions of the polarization and absorption of light in water (Boas 1881), a study that influenced his later interests in the cultural diversity of perception and knowledge. This confusion of ideas of light with ideas of knowledge is visible too in Fox Talbot's foundational description of his calotypes as 'words of light' (Fox Talbot 1996 [1839]: xxxiv), and perhaps also the visual imagination that characterized debates among Victorian archaeologists about linguistic and material knowledge, as with Pitt-Rivers's account of a gradual process of 'realistic degeneration' from 'pictographic representations . . . into phonetic characters', in which words, images and tools can be 'classed together' (Lane Fox 1875: 500, 515).

This recollection of Victorian ideas of visual knowledge in Archaeology and Photography reminds us that the archaeographic image is not just a trace or a remnant, an illustration or by-product, or even a snap-shot or artefact of past archaeo-photographic practice. In many fields, as John Berger once observed, a photograph will often be 'used tautologically' so that it 'merely repeats what is being said in words' (Berger 1976: 81). But what emerges through Archaeology is a visual modality of knowledge of the past which, reversing our conception of archaeography, I want to call *photological*.

The archaeological gesture, both interventionist and curatorial, serves not to fix an image as a ruin but to destabilize it as an object, transforming it into the very opposite of a still (Hicks 2016b). Photology draws our attention to the unstillness of the archaeological photograph as a form of knowledge of the past. This longstanding process of the destabilization and unstill nature of visual knowledge in Archaeology and Photography is one that we recognize from the hyperactive transformations of the social economy of digital images: the timelines of comments and tags, the ephemerality of transitory photo-sharing through which Snapchat emerges alongside Photoshop, the recursive nature of acts of reblogging, retweeting and the 'Facebook memory', the temporal protentions of archiving and sharing through which media gain new social lives through the open-ended classificatory impulses making new ontologies, taxonomies, folksonomies as forms of a visual archaeology of the emergent. The social media companies that produce these operations function as new kinds of archive – but on a model that combines images and knowledge in perhaps a recognizably archaeological manner.

This emergent photocentric global culture coincides, as noted above, with an unfolding of the academic discipline of Archaeology into the most recent and contemporary worlds. Part of the challenge that this presents for the mainstream literature on Photography is nicely expressed by Danny Miller as a kind of counterflow to the development of the field of Visual Anthropology:

> this movement of anthropology towards photography has been met by a counter movement from photography towards anthropology. With over a billion photographs posted every day, photography has expanded to become a ubiquitous presence. Through platforms such as Twitter and WhatsApp it has become an integral part of messaging and general communication. This storm, this deluge of photographs now saturates almost every relationship, every concern and every interest anthropologists may wish to explore, from kinship to shopping. It is hard to imagine any topic of ethnography that would not be enhanced by studying social-media photography, which shifts this from a sub-discipline to as much an integral part of ethnography as conversation.
>
> MILLER 2015: 14

Importantly, Miller questions how much is unprecedented in the anthropological study of 'photography in the age of Snapchat'. He introduces the notion of 'contemporary or ethnographic history' as a way of understanding the complex sequences and layers of images in social media, through which photography becomes 'almost analogous to language itself' (Miller 2015: 1, 14). This not-quite analogy, which perhaps we might rephrase as another Barthesian 'confusion of practice', not only reminds us of how images and texts are bound up together in a range of new and unstable configurations today, but also suggests deeper points of identification between Photography and knowledge. In this respect, Miller's analysis echoes Nick Mirzoeff's 'counterhistory of visualism' in which he observes how 'closely imbricated' words and images were throughout the modern period (Mirzoeff 2011: 15). Social media renders new forms of personhood, and perhaps even humanity (Whitehead and Wesch 2012), that complicate authorship and introduce a distribution of temporal registers. There is a blurring of the professional and the vernacular as genres (as in participant observation) and a distribution of human and nonhuman forms of life (like the products of excavation). The knowledge explosion ignited by contemporary changes in social media and global connectivity runs from 'folksonomies' and data mining to crowdsourcing and collaborative publishing, human-based computation, broadcasting and location-based and real-time web, and the many more negative effects of these technologies. These develop hand-in-hand with 'new forms of visualization and "orders of visibility"', each of which brings new 'democratic opportunities and risks' (Grau and Veigl 2013: 1, 5), as well as a vast array of visual distractions across the screens that proliferate across everyday life in so much of the West. But any case for the exceptionalism of the present is unclear.

Indeed, these contemporary transformations may reveal more longstanding processes that are inherent to Archaeology as *a discipline made possible by Photography* (the optics that lay beneath the technology for knowing the past). But how was the emergence of Archaeology from within the ongoing antiquarian tradition contingent on a mode of visualism made possible by the invention of photography in 1839, and what could that mean for Archaeology and Photography today? Any account of Visual Archaeology, and especially one that seeks to make a break from an artefactualist account of Archaeology and Photography, requires a theory of visualism. The Representational Archaeology has emphasized the historical circumstances of the development of Archaeology and Photography: that 'archaeology and photography emerged at the same time' (Shanks 2016) as 'devices of Western modernity' (Hamilakis, Anagnostopoulos and Ifantidis 2009: 285). Against this view of a purely contemporaneous and symmetrical relationship, there is a sense in which our present socio-digital moment constitutes an unfolding of longer-term patterns. This unfolding operates in quite the reverse direction from any emergence of the ruin or remnant, or a mere shift in scale or scope. Like the transformation of Archaeology and Photography for which this chapter is written, so the socio-digital landscape emerges through much more longstanding 'confusions' of knowledge and image, of then and here, of there and now, of subject and objects – *idiorrhythms* across modes of time, of knowledge, of humanity. Across two parts (two parts of an essay, for example).

The study of Archaeology and Photography must thus not confine us, through an artefactualist reading of the *photographique* as analogy or theory, to a domain of mere nonhuman ruins and remnants. *Photological* knowledge emerges through modalities of living together that have been actively confused between Archaeology and Photography since the invention of Photography in 1839 (an 'anthropological revolution', Barthes 1964b: 47). Two dimensions of causality suggest themselves. First, that in creating images that (unlike painting or sculpture) were never simply remnants of human practice *photography may have created the visual conditions for archaeological knowledge, which comes from returning, re-enacting, and thus transforming*. Photography underlies Archaeology, a discipline of visual knowledge (*photology*) of the seen and the unseen. Second, that in the archive, the museum vitrine and the lecture hall *Archaeology may have made this visual knowledge social*. ('From the point of view of a historical anthropology, the New Absolute, the mutation, the threshold, is Photography' (Barthes [2013: 71]).)

The idea of *photology* – the knowledge made through archaeology's mode of visualism – opens the possibility for a sense of Archaeology and Photography not as an interpretive endeavour but as a conjectural and disjunctive discipline: nonrepresentational, speculative, unequivocal and asymmetrical, uncontemporary not analogical, intensely reciprocal not summative, durational rather than 'multi-temporal'. Its vision is to run into the past to creating visual disturbances in which the immense potentiality of the latent image is not just to reveal the past, but to transform it in a recursive manner.

A transformation of Visual Archaeology might therefore run as follows.

First dismantle the appealing but superficial reflexivity that accompanied the idea that Photography and Archaeology as jointly engaged in documenting the same kind of object, that they are 'sharing an ontology' (Shanks and Svabo 2013).

In its place, imagine that Archaeology's visualism begets not representation through display but transformation through discovery, looking across time but not enduring unidirectionally, and that the technology of Photography was its precondition, through which it emerged from antiquarianism through a kind of Barthesian 'identification'.

If Archaeography is Archaeology and Photography confused through method not through value, then it can be truly 'analogous to ethnographic description' (Deetz 1991: 430) only insofar as it transforms human communities past and present rather than just representing nonhuman remnants. So let us imagine Photology as the modality through which Archaeology identifies with Photography, lives together, from one part to another: a Barthesian 'confusion of methods' that yields photology – knowledge of the past wrought through light.

Notes

1 'Das andre, was ich nicht hören mag, ist ein berüchtigtes »und«: die Deutschen sagen »Goethe *und* Schiller«, – ich fürchte, sie sagen »Schiller und Goethe« . . . *Kennt* man noch nicht diesen Schiller? – Es gibt noch schlimmere »und«; ich habe mit meinen eigenen Ohren, allerdings nur unter Universitäts-Professoren, gehört »Schopenhauer *und* Hartmann«'.

2 All translations from French and German in this essay (both Part One and Part Two), unless otherwise indicated, are my own.

3 'Certains auront reconnu la phrase que j'ai donnée pour titre à cette conférence : « Longtemps, je me suis couché de bonne heure. Parfois, à peine ma bougie éteinte, mes yeux se fermaient si vite, que je n'avais pas le temps de me dire: "Je m'endors". Et, une demi-heure après, la pensée qu'il était temps de chercher le sommeil m'éveillait . . . » : c'est le début de *la Recherche du temps perdu*! Est-ce à dire que je vous propose une conférence "sur" Proust ? Oui et non. Ce sera, si vous voulez bien: Proust et moi. Quelle prétention! Nietzsche ironisait sur l'usage que les Allemands faisaient de la conjonction "et": "Schopenhauer et Hartmann", railait-il. "Proust et moi" est encore plus fort. Je voudrais suggérer que, paradoxalement, la prétention tombe à partir du moment où c'est moi qui parle, et non quelque témoin: car, en disposant sur une même ligne Proust et moi-même, je ne signifie nullement que je me compare à ce grand écrivain: mais, d'une manière tout à fait différente, que *je m'identifie* à lui: confusion de pratique, non de valeur'. The English translation of the opening lines of Swann's Way used here is that of Richard Howard (See "Howard's Way', *New York Times*, 25 September 1988, p. 5). Barthes expands on

this distinction between comparing with and identifying with in the first lecture of the *The Preparation of the Novel* series (Barthes 2011: 3).

4   Bohrer's discussion here (2005: 183, 188) forms the conceptual basis of his book-length treatment of *Photography and Archaeology* (2011).

5   'Ninisme. J'appelle ainsi cette figure mythologique qui consiste à poser deux contraires et à balancer l'un par l'autre de façon à les rejeter tous deux. (Je ne veux ni de ceci, ni de cela.) C'est plutôt une figure de mythe bourgeois, car elle ressortit à une forme moderne de libéralisme. On retrouve ici la figure de la balance : le réel est d'abord réduit à des analogues; ensuite on le pèse ; enfin, l'égalité constatée, on s'en débarrasse . . . [o]n fuit le réel intolérable en le réduisant à deux contraires qui s'équilibrent dans la mesure seulement où ils sont formels, allégés de leur poids spécifique'.

6   Note that in account of *Die Zeit des Weltbides* refered to by Hamilakis *et al.*, Heidegger describes the creation of a 'world picture' as a double operation: 'durch den vorstellend-herstellenden Menschen gestellt' – 'set out by the man who represents and constructs' (Heidegger 1977 [1950]: 89).

7   'La photographie (témoin brut de « ce qui a été là »), le reportage, les expositions d'objets anciens (le succès du show Toutankhamon le montre assez), le tourisme des monuments et des lieux historiques'.

8   Shanks's account ran as follows: 'Photographs which seem to represent reality are products of discourse . . . What we take to be objective reality is rhetorically constructed . . . Microbes needed the like of Pasteur (Latour 1988). And photowork needs people and discourse. My argument is that photographs are powerful rhetorical instruments in establishing objectivity: they work as images and as products of a technique which apparently capture an objective correlative. This has led me to introduce discourse as a concept vital for understanding the social and historical production of knowledge . . . I repeat that photographic naturalism is not necessarily realistic, and our models of realism are historical.' (Shanks 1997: 81, 83, 86).

   '*None of this is news*,' exclaimed Ruth Van Dyke in her discussion of Shanks's position (Van Dyke 2006: 372), although she did point to some interesting studies of the social construction of archaeological photographs such as Joan Gero and Dolores Root's (1990) deconstruction of socio-political messages in *National Geographic* archaeological photographs – to which we might add Angela Piccini's (1996) account of the making of Celtic pasts through archaeo-historical documentary film.

9   'Dans la photographie, en effet – du moins au niveau du message littéral– , le rapport des signifiés et des signifiants n'est pas de «transformation» mais d'«enregistrement»'.

10 'Ainsi apparaît le statut particulier de l'image photographique: c'est un message sans code'.

11 Barthes' inaugural lecture for the Chair in Literary Semiology at the Collège de France was given on 7 January 1977 (Barthes 1978). His lecture series at the Collège de France in 1977–78 (Barthes 2002a) and in 1976–77 and 1977–80 (Barthes 2002b, 2003) have been published in English translations by Rosalind Krauss and Denis Hollier (Barthes 2005) and by Kate Briggs (Barthes 2011, 2013).

12 'Qu'est-ce que la photographie transmet? Par définition, la scène elle-même, le réel littéral. De l'objet à son image, il y a certes une réduction: de proportion, de perspective et de couleur. Mais cette réduction n'est à aucun moment une *transformation* (au sens mathématique du terme)'.

13 'Il a été absolument, irrécusablement présent, et cependant déjà différé'.

14 'Je voudrais une Histoire des Regards'.

15 'On ne fera jamais trop de photos'.

16 'Jeder neuen Technik, bemerkt [Lhote], liegt eine neue Optik zu Grunde'.

17 The neologism 'gastro-archaeology' was, of course, coined and expanded upon in book-length form by Glyn Daniel in 1963 in his *The Hungry Archaeologist in France* (Daniel 1963).

18 'Le regard peut alors se porter, non sans perversité, sur des choses anciennes et belles, dont le signifié est abstrait, périmé'.

19 Especially through the co-location of groups of theoretically minded geographers and archaeologists in the environment of the University of Wales, Lampeter.

20 'Je mehr die Entfernung wächst, desto klarer wird die Sicht. Mit der größtmöglichen Deutlichkeit erblickt man die winzigen Details. Es ist, als schaute man zugleich durch ein umgekehrtes Fernrohr und durch ein Mikroskop.'

21 'Jedoch bedeutet in der Optik der Geschichte – darin das Gegenbild der räumlichen – Bewegung in die Ferne Größerwerden'.

22 'Pour beaucoup de ces photos, c'était l'Histoire qui me séparait d'elles. L'Histoire, n'est-ce pas simplement ce temps où nous n'étions pas nés? Je lisais mon inexistence dans les vêtements que ma mère avait portés avant que je puisse me souvenir d'elle'.

23 'Proust cherche une form qui recueille la souffrance (il vient de la connaître, absolue, par la mort de sa mère) et la transcende'.

24 'Certes l'image n'est pas le réel ; mais elle en est du moins l'analogon parfait, et c'est précisément cette perfection analogique qui, devant le sens commun, définit la photographie'.

25 'Dans la Photographie, la présence de la chose (à un certain moment passé) n'est jamais métaphorique'.

26 'La photographie installe en effet, non pas une conscience de l'*être-là* de la chose (que toute copie pourrait provoquer), mais une conscience de l'*avoir- été-là*. Il s'agit donc d'une catégorie nouvelle de l'espace-temps : locale immédiate et temporelle antérieure ; dans la photographie il se produit une conjonction illogique entre l'*ici* et l'*autrefois*.'.

27 Tylor's paper described an apparatus for teaching optics made by him 'with the help of R. Knight' from four pieces of 'thick-piled velvety plush known as "imitation sealskin"' cut to the shapes of 'a thick plate, a prism, a convex and a concave lens and glued onto strong boards', and a runner made from four boxwood wheels on an iron axle which 'is trundled across the board'. 'Wet sand will answer equally well with the velvet, if metal wheels are used,' he added' (Tylor 1874: 158–9).

References

Baird, J.A. 2017. Framing the Past: Situating the Archaeological in Photographs. *Journal of Latin American Cultural Studies* 26(2): 165–186.

Balm, R. 2016. *Archaeology's Visual Culture: Digging and Desire*. Abingdon: Routledge.

Banks, M. 2001. *Visual Methods in Social Research*. London: Sage.

Barthes, R. 1957. *Mythologies*. Paris: Editions du Seuil.

Barthes, R. 1961. Le message photographique. *Communications* 1: 127–138.

Barthes, R. 1964a. Images, raison, deraison. Preface to *L'univers de l'Encyclopedie*. Paris: Libraires associés, pp. 11–16 (later published in English as 'The Plates of the Encyclopedia').

Barthes, R. 1964b. Rhétorique de l'image. *Communications* 4: 40–51.

Barthes, R. 1967. *Système de la Mode*. Paris: Seuil.

Barthes, R. 1968. L'effet de réel. *Communications* 11: 84–89.

Barthes, R. 1970a. La mort de l'auteur. *Mantela* V: 12–17.

Barthes, R. 1970b. *S/Z*. Paris: Seuil.

Barthes, R. 1971. Ecrivains, intellectuels, professeurs. *Tel Quel* 47: 3–18.

Barthes, R. 1977. *Fragments d'un discours amoureux*. Paris, Seuil.

Barthes, R. 1978. *Leçon. Texte de la leçon inaugurale prononcée le 7 janvier 1977 au Collège de France*. Paris: Editions du Seuil (published in English, translated by R. Howard, in 1979 as Lecture in Inauguration of the Chair of Literary Semiology, Collège de France, January 7 1977 *October* 8: 3–16.)

Barthes, R. 1980. *La Chambre claire*. Paris, Seuil/Gallimard.

Barthes, R. 1984 [1978]. Longtemps, je me suis couché de bonne heure. In *Le Bruissement de la Langue (Essais critiques IV)*. Paris: Editions du Seuil, pp. 313–325.

Barthes, R. 2002a. *Le Neutre. Cours et séminaires au Collège de France, 1977–1978* (edited by T. Clerc and E. Marty). Paris: Editions du Seuil.

Barthes, R. 2002b. *Comment vivre ensemble: simulations romanesques de quelques espaces quotidiens: notes de cours et de séminaires au Collège de France, 1976–1977* (edited by C. Coste). Paris: Editions du Seuil.

Barthes, R. 2003. *La préparation du roman, I et II: cours et séminaires au Collège de France, 1978–1979 et 1979–1980* (edited by N. Léger). Paris: Editions du Seuil.

Barthes, R. 2005. *The Neutral: Lecture Course at the Collège de France, 1977–1978* (trans. R. Krauss and D. Hollier). New York: Columbia University Press (translation of Barthes 2002a).

Barthes, R. 2011. *The Preparation of the Novel: Lecture Courses and Seminars at the Collège de France, 1978–1979 and 1979–1980* (trans. K. Briggs). New York: Columbia University Press (translation of Barthes 2003).

Barthes, R. 2013. *How to Live Together: Novelistic Simulations of Some Everyday Spaces* (trans. K. Briggs). New York: Columbia University Press (translation of Barthes 2002b).

Bateman, J. 2005. Wearing Juninho's Shirt: Record and Negotiation in Excavation Photographs. In S. Smiles and S. Moser (eds) *Envisioning the Past: Archaeology and the Image*. Oxford: Blackwell, pp. 192–203.

Bateman, J. 2006. Pictures, Ideas and Things: The Production and Currency of Archaeological Images. In M. Edgeworth (ed.) *Ethnographies of Archaeological Practice: Cultural Encounters, Material Transformations*. Lanham, MD: Altamira, pp. 68–80.

Bender, B. 1993. Introduction. Landscape: Meaning and Action. In B. Bender (ed.) *Landscape: Politics and Perspectives*. Oxford: Berg, pp. 1–17.

Bender, B. 1999. Subverting the Western Gaze: Mapping Alternative Worlds. In P.J. Ucko and R. Layton (eds) *The Archaeology and Anthropology of Landscape: Shaping Your Landscape*. London: Routledge (One World Archaeology 30), pp. 31–45.

Benjamin, W. 1972 [1927]. Moskau. In *Gesammelte Schriften Volume 4(1)*. Frankfurt: Suhrkamp, pp. 316–352.

Benjamin, W. 1972 [1936]. Pariser Brief II: Malerei und Photographie. In *Gesammelte Schriften Volume 3*. Frankfurt: Suhrkamp, pp. 495–507.

Berger, J. 1966. Arts in Society: The Uses of Photography. *New Society* 8: 582–853.

Berger, J. 1976. Drawn to That Moment. *New Society* 37: 81–82.

Blonsky, M. 1985. Introduction: The Agony of Semiotics: Reassessing the Discipline. In M. Blonsky (ed.) *On Signs*. Baltimore, MD: Johns Hopkins University Press, pp. xiii–li.

Boas, F. 1881. Beiträge zur Erkenntnis der Farbe des Wassers. Ph.D. dissertation, Kiel University.

Bohrer, F. 2005. Photography and Archaeology: The Image as Object. In S. Smiles and S. Moser (eds) *Envisioning the Past: Archaeology and the Image*. Oxford: Blackwell, pp. 180–191.

Bohrer, F. 2011. *Photography and Archaeology*. London: Reaktion.

Brassaï 1997. *Marcel Proust sous l'Emprise de la Photographie*. Paris: Gallimard.

Browne, T. 1658. *Hydriotaphia, urne-buriall, or, a discourse of the sepulchrall urnes lately found in Norfolk*. London: Printed for Henry Brome at the signe of the Gun in Ivy-lane.

Buckley, L. 2014. Photography and Photo-Elicitation after Colonialism. *Cultural Anthropology* 29 (4): 720–743.

Brusius, M. 2013. From Photographic Science to Scientific Photography: Talbot and Decipherment at the British Museum around 1850. In M. Brusius, K. Dean and C. Ramalingam (eds) *William Henry Fox Talbot: Beyond Photography*. New Haven, CT: Yale University Press, pp. 219–244.

Brusius, M. 2016. Photography's Fits and Starts: The Search for Antiquity and Its Image in Victorian Britain. *History of Photography* 40(3): 250–266.

Candea, M., J. Cook, C. Trundle and T. Yarrow (eds) 2015. *Detachment: Essays on the Limits of Relational Thinking*. Manchester: Manchester: University Press.

Carabott, P., Y. Hamilakis and E. Papargyriou 2015. Capturing the Eternal Light: Photography and Greece, photography of Greece. In P. Carabott, Y. Hamilakis and E. Papargyriou (eds) *Camera Graeca: Photographs, Narratives, Materialities*. Farnham: Ashgate, pp. 3–21.

Cochrane, A. and I. Russell 2007. Visualizing Archaeologies: A Manifesto. *Cambridge Archaeological Journal* 17(1): 3–19.

Collier, J. 1967. *Visual Anthropology: Photography as a Research Method*. New York: Holt, Rinehart and Winston (Studies in Anthropological Method).

Coombes, A.E. 2016. Photography Against the Grain: Rethinking the Colonial Archive in Kenyan Museums. *World Art* 6(1): 61–83.

Daniel, G. 1963. *The Hungry Archaeologist in France: A Travelling Guide to Caves, Graves and Good Living in the Dordogne and Brittany*. London: Faber and Faber.

Dant, T. and G. Gilloch 2002. Pictures of the Past: Benjamin and Barthes on Photography and History. *European Journal of Cultural* Studies 5 (1): 5–23.

Daston, L. and P. Galison 1992. The Image of Objectivity. *Representations* 40: 81–128.

Daston, L. and P. Galison 2007. *Objectivity*. New York: Zone Books.

de Duve, T. 1978. Time Exposure and Snapshot: The Photograph as Paradox. *October* 5: 113–125.

Deetz, J. 1991. Archaeography, Archaeology, or Archeology? *American Journal of Archaeology* 93(3): 429–435.

Desroches-Noblecourt. C. 1996. *Ramses II – la veritable histoire*. Paris : Pygmalion/G. Watelet.

Dubois, P. 2016. Trace-Image to Fiction-Image: The Unfolding of Theories of Photography from the '80s to the Present. *October* 158: 155–166.

Edwards, E. 1992. Introduction. In E. Edwards (ed.) *Anthropology and Photography, 1860–1920*. New Haven. CT: Yale University Press, pp. 3–17.

Edwards, E. 2001. *Raw Histories: Photographs, Anthropology and Museums*. Oxford: Berg.

Edwards, E. 2005. Photographs and the Sound of History. *Visual Anthropology Review* 21: 27–46.

Edwards, E. 2011. Tracing photography. In M. Banks and J. Ruby (eds) *Made to Be Seen: Perspectives on the History of Visual Anthropology*. Chicago: University of Chicago Press, pp. 159–189.

Edwards, E. and J. Hart 2004. Introduction: Photographs as Objects. In E. Edwards and J. Hart (eds) *Photographs Objects Histories: On the Materiality of Images*. London: Routledge, pp. 1–15.

Edwards, E. and C. Morton 2015. Between art and information: Towards a Collecting History of Photographs. In E. Edwards and C. Morton (eds) *Photographs, Museums, Collections*. London: Bloomsbury, pp. 3–23.

Fabian, J. 1983. *Time and the Other*. New York: Columbia University Press.

Foster, H. 1996. The Archive without Museum. *October* 77: 97–119.

Fox Talbot, W.H. 1996 [1839]. *Records of the Dawn of Photography: Talbot's Notebooks P & Q* (ed. L.J. Schaaf). Cambridge: Cambridge University Press.

Gero, J. and D. Root 1990. Public Presentations and Private Concerns: Archaeology in the Pages of National Geographic. In P. Gathercole and D. Lowenthal (eds) *The Politics of the Past*. London: Unwin Hyman, pp. 19–37.

Gibson, J.J. 1979. *The Ecological Approach to Visual Perception*. Boston: Houghton Mifflin.

Gilchrist, R. 1992. Review of Michael Shanks 'Experiencing the Past'. *Archaeological Review from Cambridge* 11(1): 188–191.

Ginzberg, C. 2001. *Wooden Eyes: Nine Reflections on Distance* (trans. M. Ryle and K. Soper). New York: Columbia University Press.

González-Ruibal, A. 2008. Time to Destroy: An Archaeology of Supermodernity. *Current Anthropology* 49(2): 247–279.

González-Ruibal, A. 2013. Reclaiming Archaeology. In A. González-Ruibal (ed.) Reclaiming Archaeology: Beyond the Tropes of Modernity. Abingdon: Routledge, pp. 1–29.

Grau, O. and T. Veigl 2013. Introduction: Imagery in the 21st Century. In O. Grau (ed.) *Imagery in the Twenty-First Century*. Cambridge, MA: MIT Press, pp. 1–17.

Griffiths, A. 1997. Knowledge and Visuality in Turn of the Century Anthropology: The Early Ethnographic Cinema of Alfred Cort Haddon and Walter Baldwin Spencer. *Visual Anthropology Review* 12(2): 18–43.

Hamilakis, Y., A. Anagnostopoulos and F. Ifantidis 2009. Postcards from the Edge of Time: Archaeology, Photography, Archaeological Ethnography (a Photo Essay). *Public Archaeology* 8 (2–3): 283–309.

Haraway, D.J. 1991. Situated Knowledges: The Science Question in Feminism and the Privilege of a Partial Perspective. In *Simians, Cyborgs, and Women: The Reinvention of Nature*. London: Free Association Books, pp. 183–201.

Harper, D. 2002. Talking about Pictures: A Case for Photo Elicitation. *Visual Studies* 17(1): 13–26.

Heidegger, M. 1977 [1950]. Die Zeit des Weltlebens. In *Holzwege (Gesamtausgabe 1, Band 5)*. Frankfurt: Vittorio Kostermann, pp. 69–96.

Hicks, D. 2003. Archaeology Unfolding: Diversity and the Loss of Isolation. *Oxford Journal of Archaeology* 22(3): 315–329.

Hicks, D. 2016a. Meshwork Fatigue. *Norwegian Archaeological Review* 49(1): 33–39.

Hicks, D. 2016b. The Temporality of the Landscape Revisited. *Norwegian Archaeological Review* 49(1): 5–22.

Hicks, D. and M.C. Beaudry 2006. Introduction: The Place of Historical Archaeology. In D. Hicks and M.C. Beaudry (eds) *The Cambridge Companion to Historical Archaeology*. Cambridge: Cambridge University Press, pp. 1–9.

Hicks, D. and L. McAtackney 2007. Landscapes as Standpoints. In D. Hicks, L. McAtackney and G. Fairclough (eds) Envisioning Landscape: Situations and Standpoints in Archaeology and Heritage. Walnut Creek, CA: Left Coast Press (One World Archaeology), pp. 13–29.

Holbraad, M. and M.A. Pedersen 2010. Planet M: The Intense Abstraction of Marilyn Strathern. *Anthropological Theory* 10(1): 1–24.

Ingold, T. 2000. *The Perception of the Environment*. London: Routledge.

Jay, M. 1993. *Downcast Eyes: The Denigration of Vision in French Twentieth Century Thought*. Berkeley: University of California Press.

Knappett, C. 2002. Photographs, Skeuomorphs and Marionettes: Some Thoughts on Mind, Agency and Object. *Journal of Material Culture* 7(1): 97–117.

Knowles, C. 2014. Negative Space: Tracing Absent Images in the National Museums Scotland's collections. In E. Edwards and S. Lien (eds) *Uncertain Images: Museums and the Work of Photographs*. Farnham: Ashgate, pp. 73–91.

Kracauer, S. 1969. *History, the Last Things Before the Last* (edited by P.O. Kristeller). Oxford: Oxford University Press.

Kracauer, S. 1993. Photography (trans. T.Y. Levin). *Critical Inquiry* 19(3): 421–436.

Kracauer, S. 2012 [1951]. The Photographic Approach. In *American Writings: Essays on Film and Popular Culture*. Berkeley: University of California Press, pp. 187–193.

Krauss, R. 1999. Reinventing the Medium. *Critical Inquiry* 25(2): 289–305.

Lane Fox, A.H. 1875. On the Evolution of Culture. *Journal of the Royal Institution* 7: 357–389. In Anon (ed.) *Notices of the Proceedings at the meetings of members of the Royal Institution of Great Britain with abstracts of the discourses delivered at evening meetings, Volume 7 (1873–1875)*. London: William Clowes and Sons, pp. 496–520.

Latour, B. 1986. Visualisation and Cognition: Drawing Things Together. *Knowledge and Society: Studies in the Sociology of Culture Past and Present*. 6: 1–40.

Latour, 1988. *The Pasteurization of France* (trans. A. Sheridan and J. Law). Cambridge, MA: Harvard University Press.

Lucas, G. 2004. *Critical Approaches to Fieldwork*. London: Routledge.

Lucas, G. 2012. *Understanding the Archaeological Record*. Cambridge: Cambridge University Press.

Lucas, G. 2015. Archaeology and Contemporaneity. *Archaeological Dialogues* 22(1): 1–15.

Lydon, J. 2016. *Photography, Humanitarianism, Empire*. London: Bloomsbury.

Mah, A. 2014. The Dereliction Tourist: Ethical Issues of Conducting Research in Areas of Industrial Ruination. *Sociological Research Online* 19(4, 13). http://www.socresonline.org.uk/19/4/13.html

Malinowski, B. 1922. *Argonauts of the Western Pacific*. London: George Routledge & Sons.

Martindale, C. 1993. *Redeeming the Text: Latin Poetry and the Hermeneutics of Reception*. Cambridge: Cambridge University Press.

Massey, D. 1991. Flexible Sexism. *Environment and Planning D: Society and Space* 9: 31–57.

Mauss, M. 1947. *Manuel d'ethnographie*. Paris: Payot.

Mead, M. 1975. Visual Anthropology in a Discipline of Words. In P. Hockings (ed.) *Principles of Visual Anthropology*. The Hague: Mouton, pp. 3–10.

Miller, D. 2015. Photography in the Age of Snapchat. London: Royal Anthropological Institute (Anthropology and Photography 1). https://www.therai.org.uk/images/stories/photography/AnthandPhotoVol1.pdf

Mirzoeff, N. 2011. *The Right to Look: A Counterhistory of Visuality*. Durham, NC: Duke University Press.

Molyneaux, B. 1997. Introduction: The Cultural Life of Images. In B. Molyneaux (ed.) *The Cultural Life of Images: Visual Representation in Archaeology*. London: Routledge, pp. 1–10.

Morton, C. 2012. Photography and the Comparative Method: The Construction of an Anthropological Archive. *Journal of the Royal Anthropological Institute* 18(2): 369–396.

Moser, S. 1992. The Visual Language of Archaeology: A Case Study of the Neanderthals. *Antiquity* 66: 831–844.

Moser, S. 2001. Archaeological Representation: The Visual Conventions for Constructing Knowledge about the Past. In I. Hodder (ed.) *Archaeological Theory Today*. Cambridge, UK, Polity Press pp. 262–283.

Moser, S. 2014. Making Expert Knowledge through the Image Connections between Antiquarian and Early Modern Scientific Illustration. *Isis* 105: 58–99.

Moser, S. and S. Smiles 2005. Introduction: The Image in Question. In S. Smiles and S. Moser (eds) *Envisioning the Past: Archaeology and the Image*. Oxford: Blackwell, pp. 1–12.

Mukerji, C. 1984. Visual Language in Science and the Exercise of Power: The Case of Cartography in Early Modern Europe. *Studies in Visual Communication* 10: 30–45.

Mullins, P. 2014. The Banality of Everyday Consumption: Collecting Contemporary Urban Materiality. *Museum Anthropology* 37(1): 46–50.

Mulvey, L. 1975. Visual Pleasure and Narrative Cinema. *Screen* 16(3): 6–18.

Nietzsche, F. 2005 [1889]. Twilight of the Idols or, How to Philosophize with a Hammer (trans. J. Norman). In A. Ridley and J. Norman (eds) *The Anti-Christ, Ecce Homo, Twilight of the Idols, and Other Writings*. Cambridge: Cambridge University Press, pp. 153–230.

Olsen, B., M. Shanks, T. Webmoor and C. Witmore 2013. *Archaeology: The Discipline of Things*. Berkeley: University of California Press.

Perry, S.E. 2011. The Archaeological Eye: Visualisation and the Institutionalisation of Academic Archaeology in London. Unpublished Ph.D. thesis, University of Southampton.

Perry, S.E. 2013. Archaeological Visualisation and the Manifestation of the Discipline: Model-making at the Institute of Archaeology, London. In B. Alberti, A. Jones and J. Pollard (eds) *Archaeology After Interpretation: Returning Materials to Archaeological Theory*. Walnut Creek, CA: Left Coast Press, pp. 281–303.

Pétursdóttir, T. and B. Olsen 2014. An Archaeology of Ruins. In *Ruin Memories. Materialities, Aesthetics and the Archaeology of the Recent Past*. Abingdon: Routledge, pp. 3–31

Piccini, A. 1996. Filming through the Mists of Time: Celtic Constructions and the Documentary. *Current Anthropology* 37: S87–S111.

Piggott, S. 1978. *Antiquity Depicted: Aspects of Archaeological Illustration*, London: Thames and Hudson.

Pink, S. 2012. Advances in Visual Methodology: An Introduction. In S. Pink (ed.)
    Advances in Visual Methodology. London: Sage, pp. 3–16.

Pinney, C. 1992. Parallel Histories of Anthropology and Photography. In E. Edwards (ed.)
    Anthropology and Photography, 1860–1920. New Haven, CT: Yale University Press,
    pp. 74–95.

Pinney, C. 2011. *Photography and Anthropology*. London: Reaktion.

Pinney, C. and N. Peterson (eds) 2003. *Photography's Other Histories*. Durham, NC:
    Duke University Press.

Proust, M. 1927. *À La Recherche du Temps Perdu XV: Le Temps Retrouvé*. Paris:
    Gallimard.

Rathje, W., M. Shanks and C. Witmore 2013. Part I: The Archaeological Imagination. In
    W. Rathje, M. Shanks and C. Witmore (eds) *Archaeology in the Making:
    Conversations through a Discipline*. Abingdon: Routledge, pp. 5–6.

Riggs, C. 2017a. Shouldering the Past: Photography, Archaeology, and Collective Effort
    at the Tomb of Tutankhamun. *History of Science* 55(3): 336–363.

Riggs, C. 2017b. Objects in the Photographic Archive: Between the Field and the
    Museum in Egyptian Archaeology. *Museum History Journal* 10(2): 140–161.

Rorty, R. 1979. *Philosophy and the Mirror of Nature*. Princeton, NJ: Princeton University
    Press.

Rose, G. 2000. *Visual Methodologies*. London: Sage.

Schiffer, M.B. 1994. Review of M. Shanks 'Experiencing the Past'. *American Antiquity*
    59(1): 158–159.

Sebald, W.G. 1995. *Die Ringe des Saturn: eine englische Wallfahrt*. Frankfurt: Eichborn
    Verlag.

Shanks, M. 1992. *Experiencing the Past*. London: Routledge.

Shanks, M. 1996. *The Classical Archaeology of Greece: Experiences of the Discipline*.
    London: Routledge.

Shanks, M. 1997. Photography and Archaeology. In B. Molyneaux (ed.) *The Cultural
    Life of Images: Visual Representation in Archaeology*. London: Routledge,
    pp. 73–107.

Shanks, M. 2007. Symmetrical Archaeology. *World Archaeology* 39(4): 589–596.

Shanks, M. 2016. Archaeography. Web page from mshanks.com recorded at archive.
    org on 19 November 2016. https://web.archive.org/web/20161119023436/http://
    www.mshanks.com/archaeology/archaeography/

Shanks, M. and M. Pearson 2001. *Theatre/Archaeology*. London: Routledge.

Shanks, M. and C. Svabo 2013. Archaeology and Photography: A Pragmatology. In
    A. González Ruibal (ed.) *Reclaiming Archaeology: Beyond the Tropes of Modernity*.
    Abingdon: Routledge, pp. 89–102.

Shanks, M. and C. Tilley 1987. *Reconstructing Archaeology*. London: Routledge.

Shanks, M. and T. Webmoor 2013. A Political Economy of Visual Media in Archaeology.
    In S. Bonde and S. Houston (eds) *Re-presenting the Past: Archaeology through
    Image and Text*. Providence, RI: Brown University, pp. 85–108.

Shanks, M. and C. Witmore 2012. Archaeology 2.0? Review of Archaeology 2.0: New
    Approaches to Communication and Collaboration [Web Book]. *Internet Archaeology*
    32. https://doi.org/10.11141/ia.32.7

Shattuck, R. 1963. *Proust's Binoculars: A Study of Memory, Time and Recognition*. New
    York: Alfred A. Knopf.

Silverman, K. 2015. *The Miracle of Analogy or The History of Photography, Part 1*.
    Stanford, CA: Stanford University Press.

Strathern, M. 1987. An Awkward Relationship: The Case of Feminism and Anthropology. *Signs* 12(2): 276–292.

Strathern, M. 1999. *Property, Substance and Effect: Anthropological Essays on Persons and Things*. London: Athlone Press.

Thomas, J. 1993. The politics of vision and archaeologies of landscape. In B. Bender (ed.) *Landscape: Politics and Perspectives*. Oxford: Berg, pp. 19–48.

Thomas, J. 2003. *Archaeology and Modernity*. London: Routledge.

Tylor, E.B. 1874. Refraction of Light Mechanically Illustrated: Questions Arising out of a Lecture at Taunton College School. *Nature* 9: 158–159.

Tylor, E.B. 1876. Dammann's Race-Photographs. *Nature* 13: 184–185.

Van Dyke, R. 2006. Seeing the Past: Visual Media in Archaeology. *American Anthropologist* 108(2): 370–384.

Viveiros de Castro, E. 2003. *And: after-dinner speech given at Anthropology and Science, the fifth Decennial Conference of the Association of Social Anthropologists of the UK and Commonwealth*. Manchester: University of Manchester.

Webmoor, T. 2013. STS, Symmetry, Archaeology. In P. Graves-Brown, R. Harrison and A. Piccini (eds) *The Oxford Handbook of the Archaeology of the Contemporary World*. Oxford: Oxford University Press, pp. 105–120.

Whitehead, N. and M. Wesch 2012. Introduction: Human No More. In N. Whitehead and M. Wesch (eds) *Human No More: Digital Subjectivities, Unhuman Subjects, and the End of Anthropology*. Boulder: University of Colorado Press, pp. 1–10.

Willerslev, R. 2007. 'To Have the World at a Distance': Reconsidering the Significance of Vision for Social Anthropology. In C. Grasseni (ed.) *Skilled Visions: Between Apprenticeship and Standards*. New York: Berghahn, pp. 23–46.

Witmore, C. 2009. Prolegomena to Open Pasts: On Archaeological Memory Practices. *Archaeologies* 5(3): 511–545.

Zheng, D. 2012. The Sideways Gaze: Roland Barthes's Travels in China. *LA Review of Books* (23 June 2012). https://lareviewofbooks.org/article/the-sideways-gaze-roland-barthess-travels-in-china/

3

'AT ANY GIVEN MOMENT': DURATION IN ARCHAEOLOGY AND PHOTOGRAPHY

Mark Knight and Lesley McFadyen

A thousand incidents arise, which seem to be cut off from those which precede them, and to be disconnected from those which follow. Discontinuous though they appear, however, in point of fact, they stand out against the continuity of a background on which they are designed, and to which indeed they owe the intervals that separate them; they are the beats of the drum which break forth here and there in the symphony. Our attention fixes on them because they interest it more, but each of them is borne by the fluid mass of our whole psychical existence. Each is only the best illuminated point of a moving zone which comprises all that we feel or think or will – all, in short, that we are at any given moment.

HENRI BERGSON 1998 [1911]: 2–3

Duration

The title of this chapter comes from the opening of Henri Bergson's book *Creative Evolution*, where he is introducing his understanding of duration. The philosophy of Bergson, his focus on duration (rather than a fixed point in time) and extent (rather than immobile sections of space), are about understanding the world in movement and this also features strongly in our archaeological research (Knight and Brudenell forthcoming; McFadyen 2006, 2008). In particular we are interested in the relationship between duration, extent and movement in archaeology and photography, and how we might understand these relations in a more critical manner if we think on the link between photography and archaeology.

This chapter considers the duration involved in the photographic process of early photography, it considers movement in several early and contemporary photographs, and it uses these to rethink the nature of archaeological evidence (Knight 2005) and the nature of movement in contemporary archaeological digital photography. Our case study from archaeology is the site of Must Farm, Cambridgeshire with its array of evidence for Bronze Age settlement (920–790 BC; Knight 2016). The architectural elements of the site include a timber causeway, and pile dwellings bounded by a palisade. However, instead of focusing on static architectural forms, we want to articulate an understanding of the site through the dynamics of building and living – architecture *as* practice, rather than architecture as an object (Hill 2003; McFadyen 2007, 2012, 2016). This way of understanding the archaeology of the built environment means that we would describe this digital composite photograph of the site (Figure 3.1) as a 'dense territory of occupation' (Rendell 2002: 5) rather than one absent of people, with the juxtaposition of things in time.

There is the making, use and unmaking of an occupied space here, and not necessarily in that sequential order. There is a scatter of fresh wood chips from the making of the ash posts, the rounded ends of ash posts inserted into the silts that make up the diagonal line of the palisade (from the top left to bottom right of the frame), and to the right of this a dense cluster of fresh wood chips from the initial construction of the pile dwellings, the criss-crossing of large worked timbers (burnt and unburnt) from the collapse of the houses, and the rims of

Figure 3.1 The palisade and pile dwelling at Must Farm, Cambridgeshire, 2006 (Cambridge Archaeological Unit).

several used pots peeking out from underneath those timbers. Extending from the riverbank are the embedded horizontal tracks of foot and hoof prints. Cutting through the riverbank (near the upper left side of the frame), and river silts (in the right-hand side of the image), are the horizontal lines of the edges of the excavation, as are the heads of several grid pegs driven into the mud. In form and line a spatial archaeology could be described from the photographs, but it is not a flat and motionless representation of past events. Around and between the so-called edges of things there is texture and depth in the photographs, there is fluidity and an extent to things, there is duration over space.

Time is not at odds, or in stages, in the archaeological evidence or in the archaeological photographs.

*

Concrete duration

Time is not usually described in terms of duration (i.e. defined by its fluidity and extent or indivisibility) in photographs. Indeed, movement in photography is more often attributed to the 'perceptual moment' of the emergence of the technologies of microtime and the tenth of a second (Canales 2009). A snapshot photograph is supposedly instantaneous and captures all, freezing a moment in time (Hamilakis and Ifantidis 2015: 138). It is able to clarify form by rendering a thing immobile (Bergson 1998 [1911]: 300). But is time frozen by, or in, the photograph, or does the image involve many events and processes (Edwards 1999; Baird this volume)? It always takes time to take a photograph. Is the metaphor of the photographic snapshot the best way to develop an understanding of the nuances of archaeological evidence (Buchli 2013: 7, 65)?

In the Spring of 1845, William Henry Fox Talbot took a photograph of The Royal Exchange, London (Figure 3.2). This is a digital copy of a salt print from a calotype negative (the negative-to-positive, multiple image paper photographic process). Important here is that the box camera would have been set up, Fox Talbot would have uncovered the lens, and then he would have looked at his pocket watch for two or three minutes before covering the lens again (Ollman 2002). During that time, movement continued. As Jimena Canales writes about Bergson, 'real movement, real change, and real events escaped between the static intervals of time used in the sciences' (Canales 2009: 17). And so a carriage wheel continues parallel to the street kerb as it follows the extension of the road into Threadneedle Street. And two people circumnavigate the railings around the statue of the Duke of Wellington. Other things and other people were moving too. But the carriage and the couple had a slower tempo as they negotiated the street and the railings. The fluidity of their moving zone is in time with the duration

Figure 3.2 The Royal Exchange, London taken by William Henry Fox Talbot in 1845 (National Science Media, London; Museum image 10314791).

of the exposure. This is the connection between Photography and Archaeology: their similar temporal extent.

This early photograph is important to archaeology because it makes us look harder at material and movement, that which is animate and inanimate. The temporal is key here, not just the spatial. There is movement between objects and parts: river sediments to ash post at Must Farm and street kerb to carriage wheel at The Royal Exchange. Fox Talbot's photograph is not a failure for having blurred form, for where there is a lack of clarity there is real movement. If we were to slow down further, the areas where we think we see form and line, dressed stone and building, would blur and move too.

*

Clock time

Both the box camera and the pocket watch are integral to the early photographic process. Time is fundamental to the creation of the image. But it is as if we have forgotten the pocket watch. It is important to think about the actual time of the

Figure 3.3 Nicolaas Henneman taking a calotype portrait of Pullen. Photograph taken by William Henry Fox Talbot, circa 1841 (National Science Media, London; Museum image 10300026).

photograph, rather than the idea of time captured by the photograph. As a material reminder, the image reproduced here as Figure 3.3 is a digital copy of another calotype by William Henry Fox Talbot from 1841 of *Nicolaas Henneman taking a calotype portrait of Pullen.* In the photograph of the taking of a photograph, Henneman is touching the box camera but he has his eyes firmly on the pocket watch.

We can learn about the moving zone and an extended duration in contemporary photography, too. In *Theater* (1993), Hiroshi Sugimoto left the shutter of his camera open for the length of the showing of a film. Time is approached as duration in his photography. The art critic Masashi Ogura writes, 'what is in question here is the way photography relates to duration as well as instantaneousness using the physical mechanism of a camera' (Ogura 2001: 7). Due to the extended duration, the details from the instantaneous images on the screen disappear and their light reveals the nuances of the material world of the cinema. Sugimoto is not interested in freezing form on the screen; he is interested in the movement it generates.

*

Movement

In 2005, Martin Newth produced a series of photographs entitled *Rush Hour – Roads*. *Rush Hour, M1 South* (Figure 3.4) was produced with an hour-long exposure time over the period of the rush hour. At first glance, the material world looks immovable as if at a point of rest (Bergson 1998 [1911]: 299). Martin Newth (pers. comm., 20 August 2017) describes this first glance as appearing to be completely empty of any traffic or people. The second glance brings 'the power to imagine the roads roaring with traffic – picking up on very small clues in the image and/or the title'.

Yet, surprisingly, form is not crystal clear. At a second glance, you look to the material, the so-called inanimate, in order to reveal the animate. If you look hard enough, take some time, you notice that these are movement images.

Figure 3.4 Rush Hour. M1 South (Martin Newth, 2005).

The car or wheel is not visible next to the painted road marking, instead a blur of movement is registered high above the white line. Unlike the William Henry Fox Talbot photograph, there is a marked difference between the fast tempo of the car and the slow duration of the photograph. The moving zone is less material, but it has a greater depth to it.

Time is not spatialized, fixed or frozen (Bergson 1998 [1911]: 303). With extended duration, photography makes more of movement, texture and depth. These photographs by Fox Talbot, Sugimoto and Newth hold to an animacy that is already there (thus not about reanimating the past).

*

Illuminated points

Eadweard Muybridge's photography is famous for supposedly capturing high-speed motion. 'His 1878 camera shutters,' Rebecca Solnit writes, 'were a triumph of engineering that made reliable exposures of a fraction of a second for the first time, a speed at which extremely rapid motion could be captured in focus rather than recorded as blurs' (Solnit 2003: 4). As Tim Cresswell describes, Muybridge's film and shutter technology could take pictures at one thousandth of a second in rapid succession (Cresswell 2006: 60). However, we would argue that it is a mistake to reach for his photographs when thinking about movement in the photographic image. Each of the examples of 'motion' in *Animal Locomotion* (Muybridge 1887) was produced by a series of cameras and a series of photographs. A collotype that is composed of 24 successive images shows this.

Twenty-four cameras were set up in a line at regular intervals and the shutters were then released consecutively. These instantaneous images are of immobile form, with each at a point of rest. Fast movement in the world can produce, through a minimum duration of exposure in photography, an immobile section of space in the photographic image. The linear and sequential arrangement of Muybridge's images, along with the numbers that count incrementally in the bottom left-hand corner of each photograph, allow the intellect to preside over actions and so picture the end-point. Bergson writes,

> So if our activity always aims at a result into which it is momentarily fitted, our perception must retain of the material world, at every moment, only a state into which it is provisionally placed.
>
> BERGSON 1998 [1911]: 300

Is movement best described by focusing on the start and end points? Or does movement operate in the intervals that separate the points – through the particular

Figure 3.5 Woman Jumping. From Eadweard Muybridge's *Animal Locomotion* (1887, plate 173) (courtesy of the Library of Congress).

qualities of its duration and extent (Aldred 2017: 91)? The aim of Muybridge's photographic work was to reveal motion through clarity of form, and yet for such form to be clear it has to be immobile. Form here is *derivative* (Canales 2009). Cresswell sees things differently 'Muybridge's images,' he argues,

> are remarkable in many ways. Most obviously they made visible the world of motion. Photography had, for a long time, been a technology that extracted stillness from the motion of the world.
>
> CRESSWELL 2006: 61

But Cresswell (2006) and Solnit (2003) mistake clarity of form for visible motion. To compound this mistake, they also write in negative terms of the blurred durations of early photography: its inability or difficulty in producing clear form. They dismiss slower readings of speed: the blur and the particular qualities of movement that it holds.

There is a second way in which the idea of movement in Muybridge's photographs is heteronomous. The apparatus of the giphoscope (i.e. the machine that makes the 'flipbook' flip) is required if we are to create the illusion of movement with these images. The movement is in the apparatus (Deleuze 1989). It is a mistake to confuse the moving object (even when photographed successively) with the act of moving. Instead, as Bergson writes, to see movement we should be 'attaching ourselves to the inner becoming of things rather than placing ourselves outside them in order to recompose their becoming artificially' (Bergson 1998 [1911]: 306). *Animal Locomotion* attempts to bring together the instant of taking the photograph and the time elapsing within the frame. But real movement involves concrete duration, and indivisible time. Muybridge's images do not give duration an absolute existence but instead put time on the same plane as space – there is no texture or depth in them, no moving zone.

*

Moving zone

We are fed up of hearing that as archaeologists we only study material remains, that the spaces we investigate are devoid of human movement and life. Archaeological evidence is often conceived in terms of residues or remains, defined by human absence rather than presence (Barrett 1988, 2006; Lucas 2012: 11-17; cf. McFadyen 2010). The assumption is that archaeological evidence is a material that has become separated from human bodies, that human processes and events have been produced and are now over and of the past. Art historian Frederick Bohrer has gone one step further: he would have it that Photography and Archaeology are both *defined by* absence (Bohrer 2011: 7).

Our view is quite the reverse.

In both Archaeology and Photography, people's relationships with things, and the practices and objects they are caught up in, are very much present. Bodies are still involved in movement. Materials go on facilitating movement (as much as having been produced by it). It is simply that the archaeologist needs to look/ think with a slower temporal register. Jeffrey Cohen put this beautifully when he wrote

> When we decelerate, imagine a deeper past, get geologic, then history becomes more eventful, richer, deeper in its strata. Modernity loses some of its lustre, prehistory loses homogeneity, and the agency of the material world becomes easier to perceive alongside that of the human.
>
> COHEN 2015: 37

Archaeological knowledge is often imagined as involving a temporal movement from dusty remains and ruined fragments (where people have been absent for a very long time) to a snapshot or trace of a particular life. Neither side of this description is accurate, because duration is inherent to archaeological evidence (cf. Lucas 2005). Far from ruins and traces, the occupied spaces of Archaeology and Photography are the product of movement and constitute a context in which future movement can occur.

We described the digital composite photograph of the Must Farm Platform as a juxtaposition of things from many times with the making, use and unmaking of an occupied space (Figure 3.1). Some have sought to understand such spaces through the idea of 'multi-temporality' in archaeo-photography (Hamilakis and Ifantidis 2015: 143). Our focus on duration leads to a very different vision. In our discussion of this photograph above, we illuminated points of a moving zone. However, time is not fixed by space. Movement is happening between other objects or parts. The architecture extends into the riverbank that is a *rodden* (the dried raised bed of a watercourse) composed of marine silts. Fish hooks and carp bone are a part of the freshwater river silts. Leaves, as well as footprints, fall in an autumnal deposition. Our understanding of these geological and ecological relationships develops in terms of how people engaged with different temporal conditions – the flooding of the Flag Fen Basin, the flow of a river channel, the season of the year (Knight 2016).

By processing the elemental in the photograph we build a framework that understands the inhuman on human terms in Archaeology. Just like Newth's photograph (Figure 3.4), movement in the archaeological evidence is less material than when seen at the pace of lived perception, but it is thicker. Archaeological evidence witnesses duration at a different scale:

> There is no form, since form is immobile and reality is movement. What is real is the continual *change of form*: form is only a snapshot view of transition.
>
> BERGSON 1998 [1911]: 302; see CANALES 2009: 186

Movement occurs as the passage of time as well as the traversal of space. For this reason the snapshot metaphor for Photography is not a good one for Archaeology. More correctly, we should be thinking about time *and* space movement, and how both of these constitute and produce context. Consider the long exposure in early photography again: it is by looking at the material, the so-called inanimate, that the animate is revealed. Time is not fixed by space. Movement happens between objects or parts. It is relationships between objects and people that are important.

*

Long exposures

'Bell Street, from High Street' (Figure 3.6a) is a carbon print of a calotype photograph by Thomas Annan taken some time between 1868 and 1877 (Annan 1977). It is one of thirty-one photographs in the series. The main focus of this photographic work is the backstreets of Glasgow. These were dark and narrow spaces, and so Annan used a large-scale camera and a wet callodian process in order to create the negatives (Mozley 1977). He would set up the camera and coat the photographic plates on the spot to produce the negatives. The process was awkward. It drew people's attention to the camera. It took time.

(Anita Mozley writes that Annan's focus was on the tenements, and that he was ambivalent to the presence of their occupants [Mozley 1977: vii], but more accurately his focus whilst taking each photograph must have been, as with Nicolaas Henneman above, on his pocket watch.)

As an experiment, we ask you to look at the photograph (Figure 3.6a) and then read the following lines from Robert Musil's *The Man Without Qualities*.

> Dark patches of pedestrian bustle formed into cloudy streams. Where stronger lines of speed transected their loose-woven hurrying, they clotted up – only to trickle on all the faster then after a few ripples regain their regular pulse-beat . . . and as a whole resembled a seething, bubbling fluid in a vessel consisting of the solid material of buildings, laws, regulations, and historical traditions.
>
> MUSIL 1953: 3

Now look at the photograph again, and mark the different kinds of blurred form. People with 'stronger lines of speed' are there as 'cloudy streams'. They are on

Figure 3.6 Bell Street, from High Street. From Thomas Annan's *Photographs of the Old Closes and Streets of Glasgow, 1868/1877* (plate 14). Overall image (a) and three details (b, c, d).

the pavement and the road, parallel to the street kerb. 'Darker patches' form where people have slowed down and stopped to talk to each other on the street corner. This is a conversation serious enough for a basket to have been put down, for it will take time. A more 'loose-woven' group talk to each other as they cross the street, whilst another play on and off the threshold of a doorway. In Figure 3.6b, one woman is doubled and repeated as she walks around the street corner and comes back to halt at a street lamp. In Figure 3.6c, the line of the street kerb not only divides the space between road and pavement (horizontally), it has depth (vertically), and provides seating space for women to halt. And the basket has been put down (it could almost already be in the museum!), but the people involved in the using and depositing of it are there, in movement. This is because of the time played out in the taking of the photograph (Figure 3.6d).

In this photograph, different kinds of moving zone are constituted through varying speeds of changing forms. Each has a different texture and depth with stronger or weaker lines, or denser to looser clusters. In this photograph, the cast of the shadow is an effect of action rather than representation. Trace is constituted from movement and a specific temporal parameter. It is not the image of something (Doane 2003). Interestingly, in the 1900 edition of Annan's work, the images were produced as photogravure plates and in that process many of the calotypes were 'tidied up' by James Craig Annan, Thomas Annan's son. (Mozley (1977: xii) writes, 'Moving figures, those ghosts who would not stand still for the photographer, are completely excised'.

*

Taking time

One difference between Archaeology and Photography is that archaeological evidence is constantly in the process of emerging. There is no lens cap that is taken off or put back on. Movement occurs in concrete duration, which is indivisible time. An archaeology in time casts light on things that are always already animate. As with Robert Musil's account of the streets of Vienna, so with this archaeological digital photograph taken at Must Farm (Figure 3.7). With the duration of exposure in early photographic technologies in mind as we consider the pace of archaeological knowledge, we can slow down our reading of the image. Human movement and life is far from absent from this archaeological evidence. The basket is an eel trap. Half of it has been uncovered in the image. The people involved in the using, depositing and uncovering this Bronze Age trap, woven from canes of willow, are there in movement. Ongoing processes are material in the river silts. In the top third of the image there is an intersection where tarpaulin, bucket and archaeologists' foot prints connect in the clods of

Figure 3.7 Eel trap in the river silts of the palaeochannel at Must Farm, Cambridgeshire, 2011 (Cambridge Archaeological Unit).

the surface mud. In the middle of the image, the archaeologists' handiwork continues where their feet do not, into the smear of mud over the eel trap. In the bottom third their handiwork moves down into the river silts where they encounter the hard edges of willow canes left where Bronze Age hands made and set the trap in the river. The trap is located deep in the petrol grey-blue river silts *and* it is positioned to catch eel in the Spring/Summer flow of a freshwater river. There are different kinds of moving zone that are constituted from the varying speeds of the changing forms. The lines made by the speed of the archaeologists' feet and hands are as strong and weak as 'dark patches' and 'cloudy streams' respectively. The hands of those that made and set the trap are less material, they are more 'loose-woven', but they have greater depth and they hold to the material. A river flows and forms through them all.

By taking time we acknowledge different kinds of form, and so we see movement.

Archaeological evidence, of which this archaeological digital photograph is one example, is defined not by absence and ruin but by presence – 'dense territories of occupation' (Rendell 2002: 5). Archaeology is not the study *of* the duration, extension and movement of other people's lives, because there is no lens cap to put on. It is a subject that is a part of those conditions as objects.

*

Conclusion

This chapter has suggested that thinking through the long exposure times that characterized early photography represents an important lens through which to understand the nature of archaeological evidence. We want to slow down the reader/viewer because it is through duration, we suggest, that we come to an understanding of movement in Photography and Archaeology.

In the uses of photography for the purposes of archaeological theory, in theorizations of a connection between photography and archaeology, critiques of time through the image of the trace or the remnant are prominent (Shanks 1992; Hamilakis 2008; Bohrer 2011; Hamilakis and Ifantidis 2013, 2015; Carabott *et al.* 2015). Interestingly, all of these works take significant inspiration from Roland Barthes' *Camera Lucida* (1981). For example, Michael Shanks places significant weight on the Barthesian concept of 'punctum', that momentary form of time that draws out greater reflection and meaning in the viewer, and can be produced from the inclusion of accidental detail in a photograph that makes the viewer take a second look and notice something else, and so it 'pierces the coherent surface of coded understanding' (Shanks 1992: 146). Philip Carabott, Yannis Hamilakis and Eleni Papargyriou foreground Barthes's (1981: 12) sense of 'disturbance' in terms of what they see as 'multi-temporality':

> Photographic objects are multi-temporal things, and they condense different times as co-existence rather than succession; for example the having-been-there of the time when the photo was taken, and the here-and-now of us viewing it, and often several other times in between.
>
> CARABOTT *et al.* 2015: 10

Thus, in their discussion of remnants of Muslim tombstones on the Acropolis, Hamilakis and Ifantidis (2015: 150) argue that in 'the combined mnemonic and temporal possibilities of both apparatuses' (archaeology and photography), 'the depiction of multi-temporal archaeological fragments is . . . enhanced by the multi-temporal affordances of the photographs themselves'.

Both the punctum-inspired account of 'significant time' or the disturbance-inspired evocation of the 'multi-temporal' are effective in disrupting the notion of archaeological photography as any straightforward documentary witness. But in their assumption that a photograph represents a trace of momentary time captured by the camera, they erase from their account the time taken in its creation. Duration cannot simply be added into these accounts because – we follow Bergson (1998 [1911]) in arguing – duration has an absolute existence.

Our image of the juxtaposition of time in archaeology and photography at Must Farm (Figure 3.1) is just as susceptible as the description by Hamilakis and Ifantidis of the mutual 'multi-temporality' of photography and archaeology is to the danger of viewing time on the same plane as space. The danger is that if time is spatialized, and there is a clear form to these multiple times, then there is no texture and depth, no 'moving zone'. It may disrupt the legacy of documentary witness in archaeological photography to detail the wood chips, timber palisade, horizontal lines of the edges of excavation at Must Farm; or the eighteenth- and nineteenth-century headstones, the nineteenth- and twentieth-century landscape, the horizontal line of the metal fence. But each instance such an account presents is detail in clear form, forging each moment of time as an incident at rest in the frame (Bergson 1998 [1911]: 2).

The limitations to such a succession of still lives appear in a blur, through Bergson and duration. There is movement, not relict stillness, in archaeological evidence and in archaeological photography. Duration, extent and movement have been overlooked in how we think about Archaeology and Photography. In the process, the animate in archaeology's materials and images has been neglected. We need not dismiss work on the legacy of surveillance (Shanks 1992) or the monumentalized moment (Hamilakis and Ifantidis 2015) in archaeological photography, or the new kinds of photographic practice in archaeology in the form of photo documentaries and photo-essays (Hamilakis and Ifantidis 2015), in asking archaeologists to linger a while longer over the evidence before us.

References

Aldred, O. 2017. Where to Go from Here: Movement and Finding the Path in the Landscape. *Landscapes* 18(1): 88–92.

Annan, T. 1977. *Photographs of the Old Closes and Streets of Glasgow 1868/1877*. New York: Dover Publications

Barrett, J. 1988. Fields of Discourse: Reconstituting a Social Archaeology. *Critique of Anthropology* 7(3): 5–16.

Barrett, J. 2006. Archaeology as the Investigation of the Contexts of Humanity. In D. Papaconstantinou (ed.) *A Critical Approach to Archaeological Practice*. Oxford: Oxbow Books, pp. 194–211.

Barthes, R. 1981. *Camera Lucida: Reflections on Photography* (trans. R. Howard). New York: Hill and Wang.

Bergson, H. 1998 [1911]. *Creative Evolution* (trans. A. Mitchell). New York: Dover.

Bohrer, F.N. 2011. *Photography and Archaeology*. London: Reaktion Books.

Buchli, V. 2013. *An Anthropology of Architecture.* London: Bloomsbury.

Canales, J. 2009. *A Tenth of a Second. A History*. Chicago: Chicago University Press.

Carabott, P., Y. Hamilakis and E. Papargyriou 2015. Capturing the Eternal Light: Photography and Greece, Photography of Greece. In P. Carabott, Y. Hamilakis and E. Papargyriou (eds) *Camera Graeca: Photographs, Narratives, Materialities*. Farnham: Ashgate, pp. 3–21

Cohen, J.J. 2015. *Stone. An Ecology of the Inhuman*. Minneapolis: University of Minnesota Press.

Cresswell, T. 2006. *On the Move: Mobility in the Modern Western World*. New York and London: Routledge.

Deleuze, G. 1989. *Cinema: The Time-Image (volume 2)* (trans. H. Tomlinson and R. Caleta). London: Continuum.

Doane, M.A. 2003. *The Emergence of Cinematic Time. Modernity, Contingency, the Archive*. Cambridge, MA: Harvard University Press.

Edwards, E. 1999. Photographs as Objects of Memory. In M. Kwint, C. Breward and L. Aynsley (eds) *Material Memories. Design and Evocation*. Oxford: Berg, pp. 221–236

Hamilakis, Y. 2008. Monumentalising Place: Archaeologists, Photographers, and the Athenian Acropolis from the Eighteenth Century to the Present. In P. Rainbird (ed.) *Monuments in the Landscape: Papers in Honour of Andrew Fleming*, Stroud: Tempus, pp. 190–198.

Hamilakis, Y. and F. Ifantidis, 2013. The Other Acropolises: Multi-temporality and the Persistence of the Past. In P. Graves-Brown, R. Harrison and A. Piccini (eds) *The Oxford Handbook of the Archaeology of the Contemporary World*, Oxford: Oxford University Press, pp. 758–781.

Hamilakis, Y. and F. Ifantidis, 2015. The Photographic and the Archaeological: The 'Other Acropolis'. In P. Carabott, Y. Hamilakis and E. Papargyriou (eds) *Camera Graeca: Photographs, Narratives, Materialities*. London: Ashgate, pp. 133–158.

Hill, J. 2003. *Actions of Architecture. Architects and Creative Users*. London: Routledge.

Knight, M. (2005) Speed. Unpublished paper presented at Theoretical Archaeology Group (TAG) 2005 Conference, University of Sheffield.

Knight, M. 2016. *An Interim Report for the Archaeological Excavation of the Must Farm Pile-Dwelling*. Cambridge: Cambridge Archaeological Unit (NHPCP 6944REC).

Knight, M. and M. Brudenell, forthcoming. *Pattern and Process: Landscape Prehistories from Whittlesey Brick Pits – The King's Dyke and Bradley Fen Excavations 1998–2004*. (CAU Flag Fen Basin Depth & Time Series – Volume 1). Cambridge: Cambridge Archaeological Unit.

Lucas, G. 2005. *The Archaeology of Time*. London: Routledge.

Lucas, G. 2012. *Understanding the Archaeological Record*. Cambridge: Cambridge University Press.

McFadyen, L. 2006. Landscape. In C. Conneller and G. Warren (eds) *Mesolithic Britain and Ireland: New Approaches*. Stroud: Tempus Publishing, pp. 121–138.

McFadyen, L. 2007. Neolithic Architecture and Participation: Practices of Making in Early Neolithic Britain. In Jonathan Last (ed.) *Beyond the Grave: New Perspectives on Barrows*. Oxford: Oxbow Books, pp. 22–29.

McFadyen, L. 2008. Temporary Spaces in the Mesolithic and Neolithic. In J. Pollard (ed.) *Prehistoric Britain*. Oxford: Blackwell, pp. 121–134.

McFadyen, L. 2010. Spaces that Were Not Densely Occupied: Questioning 'Ephemeral' Evidence. In Duncan Garrow and Thomas Yarrow (eds) *Archaeology and Anthropology: Understanding Similarity, Exploring Differences*. Oxford: Oxbow Books, pp. 40–52.

McFadyen, L. 2012. The Time It Takes to Make: Design and Use in Architecture and Archaeology. In W. Gunn and J. Donovan (eds) *Design and Anthropology*, London: Ashgate, pp. 101–120.

McFadyen, L. 2016. Immanent Architecture. In M. Bille and T. Flohr-Sorensen (eds) *Elements of Architecture: Assembling Archaeology, Atmosphere and the Performance of Building Space*. London: Routledge, pp. 53–62.

Mozley, A.V. 1977. Introduction to the Dover Edition. In T. Annan *Photographs of the Old Closes and Streets of Glasgow 1868/1877*. New York: Dover, pp. v–xii.

Musil, R. 1953. *The Man Without Qualities. Volume 1* (trans. E. Wilkins and E. Kaiser). London: Secker and Warburg.

Muybridge, E. 1887. *Animal locomotion*. Philadelphia: Photogravure Company of New York.

Ogura, M. 2001. Sugimoto, a Japanese Photographer. In *Hiroshi Sugimoto. The Hasselblad Award 2001.* Goteborg: Hasselblad Centre, pp. 7–9.

Ollman, A. 2002. Henry Talbot, The Man Who Tamed Light. In M. Gray, A. Ollman and C. McCusker (eds) *First Photographs. William Henry Fox Talbot and the Birth of Photography*. New York: Powerhouse Books, pp. 13–15.

Rendell, J. 2002. Writing Aloud. In J. Macarthur and A. Moulis. *Additions to Architectural History*. Brisbane: Society of Architectural Historians of Australia and New Zealand (CD ROM), pp. 1–19.

Shanks, M. 1992. *Experiencing the Past. On the Character of Archaeology*. London: Routledge.

Solnit, R. 2003. *Motion Studies: Eadweard Muybridge and the Technological Wild West*. London: Bloomsbury.

4

EXPOSING ARCHAEOLOGY: TIME IN ARCHAEOLOGICAL PHOTOGRAPHS

J.A. Baird

I. Introduction

It's not quite possible, on the poorly exposed contact print, to see the expression on his face. Sleeves rolled up, he has the hands of someone who has been working with the earth. The doorway was not yet fully excavated but was still half-buried in earth when the man crouched within it for a photograph taken on April 21, 1928 (Figure 4.1).

The photograph was taken by Maurice Pillet, who was then Field Director of the excavation at Dura-Europos on the Syrian Euphrates. A joint expedition by Yale University and the French Academy of Inscriptions and Letters, the excavations revealed an extensive Arsacid and Roman period urban site, first settled as a Hellenistic colony, probably in the late fourth century BCE, and occupied until its fall to the Sasanian Persians in the third century CE. One of the first monuments investigated was the main gate of the city, known by the excavators as the "Palmyrene Gate" because it faced the steppe to the west of the city, across which lay Palmyra. The photograph was taken within the main passage of the gate, at the entrance to its south tower. We don't know the man's

* Susan Matheson and Lisa Brody of the Department of Ancient Art at Yale University Art Gallery graciously provided access to the Dura-Europos archives on which this work is based. The openness with which they provide access to both the physical archive and the digital images is a model in the field. Thanks are due to Simon James for scanning the index card of photograph B22, and the same image in Pillet's photograph album. Finally, thanks are due to Lesley McFadyen and Dan Hicks for the invitation to present the papers on which this chapter is based, at the Royal Anthropological Institute Conference in 2014 and European Association of Archaeologists Conference in 2015, and for their sage edits which have much improved it.

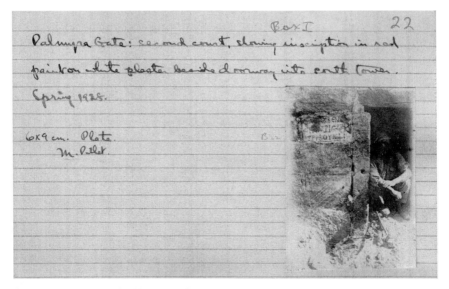

Figure 4.1 Index card with pen and pencil annotations in several hands, and a contact print showing the Palmyra Gate with inscription (see Figure 4.6) stuck to its surface from (Yale University Art Gallery Dura-Europos photograph catalogue index card B22).

name. He wears a buttoned shirt and laced shoes of the Western archaeologists, but a scarf around his head like those of the local workers. He was perhaps one of the foremen of the site who acted as intermediaries between European and American archaeologists and their local workforce (Mairs and Muratov 2015). While it is impossible to be certain, the man in the photograph might be Victor Assal, whom Pillet reports was an Armenian he hired in Deir ez Zor, to be head of the workers. Names of a few other members of staff in this season are known: in Pillet's report on the season he names only his Armenian driver, his cook, and a few others (Pillet 1928: 5–6).

The photographs and other records made in the field by Pillet and his successors were sent, for the most part, from Syria to Yale University in New Haven, Connecticut, where they remain today (Baird 2011). For this image, the negative is held in the archive, as is a contemporary print, in a photograph album compiled by Pillet. Years after the photograph was taken, a print was made from the glass negative, as the records of the excavation congealed into an archive and were organized. This new print was attached to an index card on which a short description was written. The index card sat within a photographic catalogue of the monuments of Dura-Europos which was created in the Dura archive at Yale. Over time in the archive it accumulated other annotations, written in a variety of hands and with different colours of ink. The central handwritten inscription on the index card describes the subject of the photograph from the perspective of the cataloguer in New Haven:

Palmyra gate: second court, showing inscription in red paint on white plaster beside doorway into south tower.

The man is unmentioned, serving perhaps simply as a scale just as a ranging rod might have done instead. The original caption, made directly onto the glass plate negative and later transcribed by Pillet himself into his photograph album which he submitted to Yale with his annual report as field director, was in French, but similarly described the inscription and the door:

38. Inscription et piedroit de la poterne de la tour sud. 21 Avril 1928.

On the archival sleeve in which the negative was placed years later, and then in the record in the architectural catalogue, new captions were written and the man still went unremarked: within the archive, as in the field, descriptions of this photograph note only the architecture and inscription. This photograph was not published in the excavation report: a different photograph was used, with the same door from an oblique angle more widely framed, perhaps a few steps back for the photographer (Figure 4.2).

What was this other photograph chosen for publication? Taken with the same camera it is emptied of any signs of contemporary human life. Unlike the

Figure 4.2 Entrance-gate to the south tower at Dura Europos, 21 April 1928 (Baur and Rostovtzeff 1929, Plate 2.3) (Yale University Art Gallery Dura Europos collection B20).

catalogued photograph the Greek letters of the painted inscription and the scratched inscriptions on the wall beneath it, both described in the text (Baur and Rostovtzeff, 1929: 20), cannot be discerned. A shaft of light falls across the wall and on the ground in front of the door, and the space beyond is in darkness. The relationship of the door to the larger gate structure within which it sits is not clear in the photograph, and the floor of the building wasn't yet excavated. Rather than presenting evidence of the painted inscription, or acting as a record of excavation, the photograph indexes the discovery of the site. Light cast on monuments long hidden.

A man central to an image who is undocumented in the cataloguing. A photograph of a text in which the text can't be read. Why do archaeologists take photographs, and how do they use them? Examining the photographs taken during the excavation of Dura-Europos in the 1920s and 1930s, this chapter asks two questions about the relationship between archaeological and photographic documentation. How do these practices of documentation represent and construct time? And what can we learn from what is exposed, and what is not, when they introduce what Benjamin might have called a 'tiny spark of contingency' into the archaeological archive (Benjamin, 1999: 510), perhaps preserving things never intended to be preserved?

*

II. Intended sequences

Chronology in archaeology is usually thought of in terms of absolute or relative dates. Archaeologists have constructed historical chronologies, accommodating time measurement systems like calendars within archaeological time-frames (Gardner 2012: 145–9). The idea of 'time perspectivism' has been used to explore how the discipline can make use of multiple analytical timescales, underlining the potential of archaeology as offering a distinctively long-term perspective on the past (Bailey 1983, 1987, 2007; Shanks and Tilley 1987: 118–36; Murray 1999). Meanwhile other archaeologies have drawn attention to alternative time-concepts at a more human scale, such as generational time and memory (Lucas 2005: 28–31). Conceptions of time in archaeological explanation have thus tended to privilege either long-term processes on the one hand, or culturally embedded notions of time as it was experienced and understood by people in the past, on the other. This is archaeology's problem with time. Historical/processual chronologies and material/human experiences rarely match up with each other. Photography has a parallel problem with time. Is a photograph a frozen or arrested moment, or does the image involve many events and processes that both precede and post-date that moment (Edwards 1999: 222)?

For archaeological photographs, these parallel problems with time are intertwined. Do we look through the photograph to see archaeology, or do we understand the photograph as a material object in its own right (Bohrer 2011: 26)? What do archaeological photographs reveal about the nature of archaeological time?

Archaeology and photography share a common interest in how something of the past can exist in the present. In both cases that existence is in a form that is fundamentally different from the past. And the form conceals that difference. Captions may serve to simplify periodization. The absence of the photographer and photographic equipment serves to reduce any sense of subjectivity. Cropping an image frames and flattens out the material dimensions of photographic practice. Reproduction naturalizes perspective as if the hovering view of the field could achieve omniscience, simultaneously revealing the plan and the interior.

As a genre, the archaeological record shot makes use of precise technologies of staging to present a moment that is doubly frozen: both in the instant of taking a photograph, and in the creation of a particular archaeological time within the frame (Baird 2011: 429). Archaeologists' labour works to remove any direct trace of itself from the frame. The scene is trowelled or swept clean, devoid of loose soil or other mess. Some archaeologists have discussed the sense of disenfranchisement that accompanies their removal from within that frame, as they sweep away their own footprints. As Lesley McFadyen and her colleagues asked two decades ago, 'Why are we so professional about creating an archaeology devoid of us?' (McFadyen *et al.* 1997, quoted by Lucas 2001: 13).

Figure 4.3 Houses in block G1 of Dura-Europos, 1931–1932 (Rostovtzeff *et al.* 1944, plate 12) (Yale University Art Gallery Dura-Europos collection E139).

Excavators and the mess of their field practices are not all that is erased. There's no trace of the camera either. And the photographic moment is reduced to another kind of archaeological past. Take the example of a photograph of excavated houses near the centre of the site at block G1 of Dura-Europos, taken in 1931–1932 and published in the Dura report in 1944 (Figure 4.3) (Rostovtzeff *et al*. 1944). The structures are plastered mudbrick courtyard houses, typical of the site in the Arsacid and Roman periods, in the central part of the city known in the publications as 'the agora' (Baird, 2014). Emptied of life, some of the objects relating to the houses' inhabitation were tidily moved to one side of a room, and against the rear wall of the courtyard some of these can be seen, including a larger mortar and drums of a column. But the dust from the excavator's curation of this image is not entirely settled. The earthen floor of the house is darker where it is still damp from being freshly cleaned, in those areas of shadow where it hasn't had time yet to dry out in the Syrian sun. Just visible in the upper right hand corner of the image a site worker has entered the scene, pushing a cart of earth along rail tracks, as the photographer, balanced atop a ladder or spoil heap, framed the shot. The photograph presents 'Third-Century Roman Dura' and in doing so a distinctive form of objectivity in archaeology, in which the photograph 'translates the site at a particular moment or moments in time', renders the photographic moment as an objectified past (Lucas 2012: 243; cf. Bohrer, 2011: 22–3).

Thus, far from simply objectively recording archaeological remains the methodological conventions of archaeological photographs are themselves productive of time. As early as 1904, William Flinders Petrie's *Methods and Aims in Archaeology* began to set out guidelines for creating archaeological images, including using sand packed into joints and gaps to facilitate better contrast in photographs of reliefs or stone walls (Flinders Petrie 1904: 76). By the 1950s, Maurice 'Cookie' Cookson's *Photography for Archaeologists* described how to prepare an archaeological site for photography, carefully brushing a stone wall in the moment before a photograph was taken to make it 'sparkle', and removing 'all irrelevancies and "clutter"'. 'Exclude or move the site desk and the human figure that is just visible on the edge of the picture', Cookson wrote, in instructions that aimed to remove any contemporary trace of human life from the frame (Cookson 1954: 48, 52, 58).

In these respects the archaeological photograph is less a form of temporal documentation than it is an intervention in time. Like Jeff Wall's 1993 photograph 'Sudden Gust of Wind (after Hokusai)', it seems at first glance to capture a specific instance, but has qualities – qualities which it perhaps shares with painting – which reveal it to be a staged and manipulated tableau (Mulvey 2000: 148). As an event, the archaeological photograph invokes the idea of salvaging or preserving by record (Edwards 2001). The potential for future loss or destruction is anticipated in the same moment as a future archaeological viewer is imagined as the image's implicit audience. Acts of cleaning and tidying the site are knowledge practices that transform interpretation into physical enaction, and photographs make those tableaus real. (Just like that apocryphal tale of the student archaeologist who dug a series of post-holes that turned out to be nothing but the shadow of their own head.)

Discussing photographs as 'objects of memory', Elizabeth Edwards (1999: 230) turns to Roland Barthes's account of how in photography 'the there-then becomes the here-now' (Barthes 1977: 44). Just like this, the archaeological photograph does not freeze time, but operates so as to extend the archaeological process through which the past comes to exist in the present time. But the archaeological photograph simultaneously reverses this formula. A photograph of a gate in the city wall near block M7–8 when unblocked during the American and French excavations at Dura-Europos in 1933–34, shows one worker posed within it and others are visible working just beyond (Figure 4.4). Through photography the excavation site became a place fixed in the Roman past. The here-now becomes the there-then.

But archaeological photographs can intervene with time not just through presence and absence, but through making cross-temporal analogies. While the Dura-Europos excavations were proceeding, pioneering aerial archaeologist Antoine Poidebard was recording supposedly Roman remains across the

Figure 4.4 Gate in the city wall at Dura-Europos, near block M7–8 when unblocked, with a posed site worker and others visible working beyond, 1933–1934 (Scan of film negative, Yale University Art Gallery Dura-Europos collection G36B).

Mandate territories of Syria and Lebanon (Poidebard 1934). As Daniela Helbig has argued, archaeological photography came to be a central tool in presenting historical precedent for French occupation. Through ascribing remains to the Roman past, Western hegemony was inscribed across the landscape (Helbig 2016: 287). Such cross-temporal analogies also operated at human scales, as well as across the landscape. A particular compression of ancient and modern occurred when carefully dressed and posed local people were placed within the frame of the archaeological photograph. Katerina Zacheria has observed how in the same (inter-war) period in Greece, photographs commissioned by the Greek state from the Greek Elli Sougioultzoglou-Seraidari (widely known as 'Nelly's') for government-published tourist literature presented 'parallelisms' in which contemporary Greeks were posed alongside ancient monuments to visually construct a narrative of continuity for the Greek people (Zacharia 2015: 238). Such posing of contemporary people in 'ancient' contexts is evident in archaeological photographs in a variety of colonial contexts, from Syria under the French Mandate to early twentieth-century Latin America and India under the Raj (Baird 2011: 430, 2017; Guha 2012). One similar instance from Dura, taken in 1932–1933 (Figure 4.5), poses a horse draped in ancient horse armour, and one of the workers who holds up an ancient shield, made of wood and rawhide, excavated at the site (James 2004: 129–34, 186–7). Here, the use of the camera to situate contemporary people in archaeological contexts serves not only to

Figure 4.5 Site worker at Dura-Europos posed holding an ancient shield with a horse wearing ancient barding, 1932–1933 (Scan of film negative, Yale University Art Gallery Dura-Europos collection FVII33).

'explain' archaeological features, but also to perform a retrojection of local populations, presenting them as if they are living in the past. Equally, that past is itself an artifice, whether in the form of an Orientalist illusion or, in Latin America, a fantasy of Manifest Destiny, enacted throughout archaeological sites.

Analogical artifice is a central element of the aesthetic of archaeological photography, and is redoubled when we look back at historic photographs in a digital age. As Margaret Iversen has put it, the 'resurrection of a past reality in the midst of a present simulacral world' means that 'it's not just the photographed object-world that is from the past; the technology itself is already passé' (Baetens *et al.* 2007: 167). The nostalgic encounter with scenes of past excavations combines with the outdated technology of early field photography and the retrojecting exoticism consciously framed by the archaeologist-photographer to build a sense of the uncanny.

Let us return to the photograph with which this discussion began (Figure 4.6). The label made by the archaeologists in the field records the intended subject of the photograph: the painted inscription. That inscription has since vanished to the naked eye, eroded by the winds and weather of the steppe. The man in the photograph must now, too, be dead. As an archaeological photograph that survives from the disciplinary past, it is by no means a frozen moment of time: April 21, 1928. It endures. In his discussions of archaeology and duration, Laurent Olivier has explored how the material present is 'made up of an accumulation of all previous states whose successions have built this present "as it is now"'. In this view, the present is being made up of 'all the durations of the past that have been recorded in matter' (Olivier 2001: 66; cf. Olivier 2011). But what times are recorded here, even if the photograph has successfully preserved in its record that which has now disappeared?

At the time, the inscription was catalogued as a historical text (text R1 in Baur and Rostovtzeff 1929: 32), transcribed from part of a physical monument to Greek text on a page in an archaeological report. It is through such transcribed texts that ancient inscriptions are most often encountered and studied. The photograph of the inscription represents another kind of historical text, which has in part the status of a moment. The original text had been carefully painted in red onto white plaster, a process which itself was a dedication, something perhaps even spoken as it was made, and repeated (out loud or in silence) when it was read again and again, in this high-traffic position within the city gate (Baird 2016: 17). The act of inscription was a performance in which its dedicator, a member of the military, was permitted to leave a mark on this place. The text was not an archive but a lived part of the urban environment. It did not simply last: it was from the outset dated to its moment of making. It was re-read and noticed for centuries, before becoming obscured, and then revealed again, and now erased. It is reproduced here, twice (Figures 4.1 and 4.6 above) from a digital form, scanned directly from the negative (as the other photographs in this chapter

Figure 4.6 The Palmyrene Gate, Dura-Europos. 'Inscription et piedroit de la poterne de la tour sud, 21 April 1928'. The number 38 is the sequence in the album which accompanied Pillet's field report on his excavations, where this image was used to illustrate the construction of the door within the gate (Pillet 1928: 37) (Yale University Art Gallery Dura-Europos collection B22).

have been). In doing so, parts of the accumulated texts and additional meanings of the archive are erased (Baird and McFadyen, 2014; Guha 2013: 184). Only two of its many instantiations within the archive are invoked. Just as the man, posed as he was in April 1928, is absent from the published photograph (Figure 4.2), so in each reproduction there are archival absences and survivals, captions and contexts lost and re-made.

The archaeological photograph has the indexical quality of a trace insofar as it was formed as light from archaeological materials hit the film in the duration of exposure. That light still persists in digital form and in subsequent reproductions. Whether in the stone wall or the man's face in shadow, the photograph's presence documents not so much the ancient past as it does the place of the Western archaeological project in the colonial project: the seizure of the Roman past by Western archaeologists who wrought the ancient Middle East as European. Events in the history of archaeology whose repercussions haunt us today. The man was intended to act as scale and as part of an archaeological aesthetic, but can be read anew in terms of power relations and labour.

Archaeological photography is concerned with the creation of both time and timelessness; its photographs operate as both evidence and aesthetic. For example, later in the 1928–1929 field season, when the Palmyrene gate had been more fully excavated and the door revealed, a second man stood against the adjacent wall to be photographed (Figure 4.7). The archaeologists' chalked

Figure 4.7 South interior wall of the Palmyrene Gate in 1928–1929. The doorway shown in Figures 4.1 and 4.6 is just out of the shot, its frame and inscription visible to the right. Just above the painted inscription the chalk label 'J16' is visible, part of the recording of the texts that lined the passageway of the gate. The name of the worker standing on the left is not recorded (Yale University Art Gallery, Dura-Europos Collection, scan of glass negative B27).

numbers – *J16, J17* – are shown next to the painted and scratched texts on the ancient monument. Like the human figure, these chalkmarks are reduced to a kind of timelessness that the aesthetic that guides archaeological photography creates (cf. Papalexandrou and Mauzy 2003). This timelessness is intended both to exist outside of the present, and to hold some form of transcendent value through aura, antiquity and authenticity. Other devices through which archaeological photography creates timelessness include the picturesque borrowed from painting, the unpeopled landscape that is ripe for the possession, a landscape with just enough vegetation to look uninhabited by modernity.

The timelessness of archaeological photography can also be an effect of the isolation of objects from contexts. Roger Fenton's many photographs of cuneiform tablets in the British Museum taken in the 1850s are a key set of early photographs. From the outset there was a focus in archaeological photography upon texts, both as discoveries and as objects of scientific analysis through photographic reproduction (Brusius 2009; Bohrer 2011: 33). Fenton's photographs of the tablets decontexualize them; they form an inventory and the photographs operate as a kind of proxy for the original archaeological objects. But Fenton's attempts to make photographic recording of objects part of normal museum practice were met with opposition, and abandoned (Falconer and Hide 2009: 7–8).

The place of timelessness in archaeological photography has changed since the 1920s (Figure 4.8). Take for example the series of four photographs of one inscribed altar from Dura, dedicated to the Palmyrene god Iarhibol. Here we can identify shifts in concerns, even within a solidly documentary paradigm. The first image (Figure 4.8a) was made in the field, with the later photographs taken once the altar had been sent to America. The most recent image, from 2011 (Figure 4.8d), is the most enigmatic and uncanny, perhaps the most timeless. Made by Yale University Art Gallery, it is the idea of the object as a singular work of fine art that informs how its timelessness is evoked. It is not that art museum

Figure 4.8 Four photographs (a–d) of an altar (Yale University Art Gallery 1929.385) dedicated to the Palmyrene god Iarhibol with a Greek inscription, found near the 'temple of Bel' at Dura-Europos. Photographs taken in (a) 1929, (b) 1930, (c) 1990s, and (d) 2011: first in the field, and then at Yale (Yale University Art Gallery, Dura-Europos Collection).

curators do not have a concern with context (Hoffman and Brody 2011), but nevertheless the photograph is provided with a neutral background that is reminiscent of an auction catalogue; such backgrounds and framing are frequently used to transform the messy, fragmentary, stuff of archaeology into singular works of fine art. In the 1931 publication of the same altar, the background is completely painted out in black and the altar similarly hovers in space, creating an image that can be read as authorless, although the background being painted out by hand on the negative in fact meant that it was heavily manipulated. Inscriptions were, for the excavators, the most prized archaeological objects found at Dura. By the 1920s the photography of inscriptions was an established part of archaeological recording practices. The photographs of inscriptions were posted back to the scientific directors in Paris and New Haven, Franz Cumont and Mikhail Rostovtzeff. A transcription would have communicated the text, but photographs were preferred for their distinctive evidentiary and aesthetic qualities. The photographs show how the same object can become less contextualized and more timeless via different photographic backgrounds and framing.

None of the images of the altar provide any sense of the altar's function – the bowl for libations at the top – or the context where it was found, just beside the entrance to the tower of 'the Palmyrene temple' (now known as the 'temple of Bel') beside another, uninscribed, altar (Baur and Rostovtzeff 1931: 90–1). Instead, all the photographs serve to isolate and detemporalize the object – rather than documenting its multi-temporality, or its use in deposition (see discussion by Hamilakis and Ifantidis 2015: 143–52). The very fact the same object is repeatedly re-photographed demonstrate shifting concerns, but none of these interventions bring it together with other, related, objects from this building, or highlight its use.

The Dura excavators almost never photographed artefacts in context, but instead recorded them removed from the structures and strata in which they were found, arranged according to the way in which they were understood (typologically and by material, rather than as assemblages). One example is a photograph taken during the 1932–1933 field season of a small ceramic pitcher (Figure 4.9). This object has been photographed as an *objet d'art* in a manner that recalls still life painting, with objects carefully arranged in visual appealing groups (Baird 2011: 441). In this way, the photographic practice treated artefacts as a different category of archaeological evidence from *in-situ* architecture, which itself demanded buildings be swept clean of objects or loose architectural fragments.

Both in the photographs that include the workers at Dura, and in the object photographs, there is a tension between their indexicality and their aesthetic arrangement, that is, between their status as records or evidence and their status as beautiful images. The pitcher was not the type of object that the Yale team usually recorded in any detail, as they were more concerned with imports and high-status items but it became the repeated subject of photographs

Figure 4.9 Ceramic pitcher from Dura-Europos, 1932–1933 field season (Yale University Art Gallery, Dura-Europos Collection, scan of film negative no. FVIII26).

(FVIII25, FVIII26, Dam177, taken in the field and later in the Damascus Museum, where it went in the *partage*). It wasn't the pitcher or its archaeological context that was at stake in the photograph. The resulting image became more important than its referent. The value lay not in the object itself, or what evidence of the ancient world it might represent, but instead in the aesthetically pleasing image it was used to create. In Classical Archaeology the boundary between scientific and aesthetic images remained as blurry as the distinction between archaeology and art. This was not because Classical archaeologists of the period did not think of their enterprise as being amongst emerging scientific fields – they did – but the artistic value they sought in the objects consistently transcended any notional scientific objectives. In this case, the particular pitcher was never catalogued. It may have been issued with a field number, but this wasn't noted in the photograph records.

Archaeological photography employs techniques of objectivity that seek to produce images that are more than just representations. The image in Figure 4.9 is made to be understood as a pitcher, not as a photograph of a pitcher. Archaeological photographs efface the decision-making of the photographer, so that the viewer looks *through* the image to see the archaeology in the frame, past the acts of framing. In the process, archaeological photographs become not just objective representations or subjective authored documents, but ongoing *sites of encounter between people and things*, which endure from when the photographs are made, forward to the times when they are viewed (Edwards and Morton 2009: 7), and back into the archaeological past. In this respect, archaeological photography is always a temporal intervention. Archaeology, of all disciplines, should be able to develop a more sophisticated understanding of the multiple temporalities in a photograph and the ways in which context and materiality can shape their uses. Instead, an archaeological aesthetic has trained us to pay no attention to the man standing in the frame.

*

III. Unintended consequences

Just as the man in the photograph at the start of this essay (Figure 4.1) reveals unintended histories, so too can other archaeological photographs be revealing in ways that were not intentional.

> No matter how artful the photographer, no matter how carefully posed his subject, the beholder feels an irresistible urge to search such a picture for the tiny spark of contingency, of the here and now, with which reality has (so to speak) seared the subject, to find the inconspicuous spot where in the immediacy of that long-forgotten moment the future nests so eloquently that we, looking back, may rediscover it.
>
> BENJAMIN 1999: 510

In his classic discussion of his photograph of the Entrance Gateway to the Queen's College, Oxford in *The Pencil of Nature* (1844), William Fox Talbot noted that one of the values of photography for inscriptions was that the image might capture information not noticed at the moment of photography – including the hour of the day.

> It frequently happens, moreover – and this is one of the charms of photography – that the operator himself discovers on examination, perhaps long afterwards, that he has depicted many things he had no notion of at the time. Sometimes

inscriptions and dates are found upon the buildings, or printed placards most irrelevant, are discovered upon their walls: sometimes a distant dial-plate is seen, and upon it – unconsciously recorded – the hour of the day at which the view was taken.

TALBOT 1844: 13

Photography's ability to capture the unseen in ways that other forms of recording cannot is one key point of overlap with Archaeology. But while Talbot was keen to emphasize photography's potential for scientific documentation, what went unnoticed might just as well include what was not meant to be included: the slipped hand, the accidental shadow, or other mistakes. So far, this essay has discussed the intentional effects of archaeological photography as a technology. But another principle element of the unseen, shared between Photography and Archaeology, is the unintended. How then to understand those aspects of the archaeological photograph?

Let us return to the question of what the archaeological photograph seeks to document. So, what is a mistake in an archaeological photograph? One of the tensions in archaeological photography has been, in broad terms, that between indexicality and the aesthetic: the tension between a good archaeological photograph and a good image. For instance, there is the widely known case of John Henry Haynes' at Nippur in 1889–1890 (Ousterhout 2011), where as has been noted the photographs provided no view into the trenches themselves: these were successful *images* but poor archaeological *records* (Bohrer, 2011: 50–1). So too, an image from Dura in the 1930–1931 documents the fieldwork itself, rather than the detail of archaeological remains recovered (Figure 4.10), which begs the question: what are these really images *of*? The temporal and contextual contingencies of photographs allow multiple answers, allow them to be read against the grain: a photograph intended in the 1920s to be of an inscription might now be of an archaeological worker. An accidentally captured shadow might turn a photographic mistake into a document of human relationships on archaeological site, relationships otherwise undocumented.

Perhaps the most common unintended dimension of archaeological photographs is when the presence of the photographer is made visible – through a boot within the frame, or a shadow cast across an excavated surface. Shadows that reveal the presence of the camera and the photographer are relatively frequent in the Dura archive (Figure 4.11, cf. Baird 2011: figure 12). These images were never used in the publication, perhaps for that reason: they were considered mistakes and drew attention to the production of the image rather than the archaeology. But they reveal the process of fieldwork in a different way to the posed oblique shot of the trench, indicating both human relationships and the process of the making of the image. The Yale/French Academy photographers behind the camera are exposed for a moment, their outlines entering the archive

Figure 4.10 Site of a house in block D5 under excavation at Dura-Europos, 1930–1931 field season. The circular depression caused by the collapse of the house's cistern had led excavators to believe there might have been a round temple structure here, which probably why the photograph was taken (Yale University Art Gallery, Dura-Europos Collection, scan of glass negative no. d128).

alongside the local men whose bodies are used as scales. For another, we restore those doing the real work at the site to a place in the narrative, even if we cannot restore their names (Shepherd 2003; Quirke 2010), and allow reflection on photography and its implications for relationships of labour and power at archaeological sites (Morgan 2016).

The shadow of the photographer gives some sense of what the boy sees above him. In this respect, it might help us to understand the archaeological project from a different perspective. The photographer's shadow inadvertently reveals a dynamic of power through which the archaeological project was not only a form of colonial soft power in the inter-war Middle East (Gillot 2010), but itself functioned as a type of colonial violence – a violence usually discussed in more direct relation to the state and the military (Neep 2012). The global spectacle of the destruction of archaeological sites in the Middle East by ISIS is, as Ömür Harmanşah (Harmanşah 2015), a military policy that has significant relationships not just with the history of iconoclasm but with the kind of 'image wars' that Bruno Latour has called *Iconoclash* (Latour 2002). Part of the power of these acts lies in inverting the formula of Euro-colonial archaeology: destroying the Roman remains which were designated as valuable to the West

Figure 4.11 A boy standing as a scale in an excavated Roman shop in block B8, at Dura-Europos, 1930–1931 field season. The shadow of the photographer standing above him is visible to the left (Yale University Art Gallery, Dura-Europos Collection, scan of glass negative no. D118).

from the time of the earliest archaeological expeditions. Acts of looting and destruction at sites such as Palmyra and Dura are informed by images of what is known and valued, and driven by the status of objects as an economic resource that can be extracted from the ground (cf. Casana and Panahipour 2014; Casana 2015).

Other things not meant to be seen were those that would have been painted out of negatives in order to transform artefacts from archaeological finds into *objets d'art* (as the altar of Iarhibol, discussed above). The staging was not meant to be revealed, but in some photographs a hand holding up the backdrop is inadvertently photographed. Other negatives were not altered, perhaps because the images were not needed for publication: in one such example, a young archaeological worker carefully holds up a ceramic vessel, with his hand behind the object so that he, with the background, could be easily painted out of the image (Figure 4.12). Some stories were literally painted over. But in cases such as this, in the backgrounds never intended to be seen, more of the asymmetrical powers and labours of archaeology become visible.

*

Figure 4.12 A man holds up a ceramic vessel found in the necropolis,1934–1935. He holds it from behind so that he can be removed with the rest of the background from the negative (Yale University Art Gallery, Dura-Europos Collection, scan of film negative no. H363b).

IV. Conclusions

Reconsidering archaeological photographs involves reconsidering archaeological knowledge. Photographs extend the contexts of the production of this knowledge beyond the moment of excavation, both through controlled duration and unintentionality. Archaeological photographs blur the boundaries between documentation and interpretation.

At Dura, therefore, taking photographs is a means not just of recording the archaeological site, but of creating it, often through unexpected consequences. As a technology, archaeological photography intervened actively with time and timelessness, recording both the deeper past and the present moment of fieldwork but in slips of the hand and awkward shadows, the camera always created more documentation and knowledge than was intended, foreshadowing future political dimensions of archaeology in this region. Thus, archaeology's photographic archives represent not just 'legacy data' of past excavations, but also provide opportunities for us to think through stories that were not meant to be told.

As material archives – wooden drawers of index cards with their affixed prints (Figure 4.1) – are increasingly digitized as searchable online collections, these characteristics of archaeological photographs come to the fore. Archaeological photographs are becoming less accessible as objects whilst they become more accessible as images. The digital frame may crop out information, such as that on the index card or photograph album, unless curatorial care is taken to record the archaeological archive as a material practice itself (Figure 4.6). And new cross-temporal connections, both intended and unexpected, emerge. On the one hand, the present moment is one in which alternative forms of archaeological photography might be imagined – a 'third kind' of archaeological photography that lies somewhere 'between artwork and ethnographic commentary or intervention' (Hamilakis *et al.* 2009), or that disrupts the expected archaeological photograph to highlight the multi-temporality of sites (Hamilakis and Ifantidis 2015). But on the other hand, the duty to undertake the more time-consuming primary archival work on the archives of nineteenth- and twentieth-century archaeology is a pressing one, in which the inherent qualities of the archaeological photograph, as both temporal intervention and unstable consequence, represent a powerful tool with which we can interrogate the politics of archaeology, and its legacies.

References

Baetens, J., D. Costello, J. Elkins, J. Friday, M. Iversen, S. Kriebel, M. Olin, G. Smith and J. Snyder 2007. The Art Seminar. In J. Elkins (ed.) *Photography Theory*. London: Routledge: 129–202.

Bailey, G. 1983. Concepts of Time in Quaternary Prehistory. *Annual Review of Anthropology* 12: 165–192.

Bailey, G. 1987. Breaking the Time Barrier. *Archaeological Review from Cambridge* 6: 5–20.

Bailey, G. 2007. Time Perspectives, Palimpsests and the Archaeology of Time. *Journal of Anthropological Archaeology* 26: 198–223.

Baird, J.A. 2011. Photographing Dura-Europos, 1928–1937. An Archaeology of the Archive. *American Journal of Archaeology* 115: 427–446.

Baird, J.A. 2014. *The Inner Lives of Ancient Houses: An Archaeology of Dura-Europos*. Oxford: Oxford University Press.

Baird, J.A. 2016. Private Graffiti? Scratching the Walls of Houses at Dura-Europos. In R. Benefiel and P. Keegan (eds) *Inscriptions in Private Places*. Leiden: Brill Studies in Greek and Roman Epigraphy, pp. 13–31.

Baird, J.A. 2017. Framing the Past: Situating the Archaeological in Photographs. *Journal of Latin American Cultural Studies* 25.

Baird, J.A. and L. McFadyen 2014. Towards an Archaeology of Archaeological Archives. *Archaeological Review from Cambridge* 29, 14–32.

Barthes, R. 1977. *Image Music Text.* Fontana Press, London.

Baur, P.V.C. and M.I. Rostovtzeff (eds) 1929. *The Excavations at Dura-Europos conducted by Yale University and the French Academy of Inscriptions and Letters. Preliminary Report of First Season of Work, Spring 1928*. New Haven, CT: Yale University Press.

Baur, P.V.C. and M.I. Rostovtzeff (eds) 1931. *The Excavations at Dura-Europos Conducted by Yale University and the French Academy of Inscriptions and Letters. Preliminary Report of Second Season on Work, October 1928-April 1929*. New Haven, CT: Yale University Press.

Benjamin, W. 1999. *Selected Writings*. Vol. *2, part 3, 1931–1934*. Edited by M.W. Jennings, H. Eiland, and G. Smith. Cambridge, MA: Belknap Press of Harvard University Press.

Bohrer, F.N. 2011. *Photography and Archaeology*. London: Reaktion.

Brusius, M. 2009. Inscriptions in a Double Sense: The Biography of an Early Scientific Photograph of Script. *Nuncius* 24: 367–392.

Casana, J. 2015. Satellite Imagery-Based Analysis of Archaeological Looting in Syria. *Near Eastern Archaeology* 78: 142–152.

Casana, J. and M. Panahipour 2014. Satellite-Based Monitoring of Looting and Damage to Archaeological Sites in Syria. *Journal of Eastern Mediterranean Archaeology and Heritage Studies* 2: 128–151.

Cookson, M.B. 1954. *Photography for Archaeologists*. London: Max Parrish.

Edwards, E. 1999. Photographs as Objects of Memory. In M. Kwint, C. Breward, and J. Aynsley (eds) *Material Memories: Design and Evocation*. Oxford: Berg, pp. 221–236.

Edwards, E. 2001. *Raw Histories: Photographs, Anthropology and Museums*. Oxford: Berg.

Edwards, E. and C. Morton 2009. Introduction, in E. Edwards and C. Morton (eds) *Photography, Anthropology and History: Expanding the Frame*. Farnham: Ashgate, pp. 1–24.

Falconer, J. and L. Hide 2009. *Points of View: Capturing the 19th Century in Photographs*. London: British Library.

Flinders Petrie, W.M. 1904. *Methods and Aims in Archaeology*. London: Macmillan.

Gardner, A. 2012. Time and Empire in the Roman World. *Journal of Social Archaeology* 12: 145–166.

Gillot, L. 2010. Towards a Socio-Political History of Archaeology in the Middle East: The Development of Archaeological Practice and Its Impacts on Local Communities in Syria. *Bulletin of the History of Archaeology* 20(1): 4–16.

Guha, S. 2012. Visual Histories, Photography and Archaeological Knowledge. In *Depth of Field: Photography as Art and Practice in India*. New Delhi: Lalit Kala Akademi, pp. 29–39.

Guha, S. 2013. Beyond Representations: Photographs in Archaeological Knowledge. *Complutum* 24: 173–188.

Hamilakis, Y., A. Anagnostopoulos and F. Ifantidis 2009. Postcards from the Edge of Time: Archaeology, Photography, Archaeological Ethnography (a photo-essay). *Public Archaeology* 8: 283–309.

Hamilakis, Y. and F. Ifantidis 2015. The Photographic and the Archaeological: the 'Other Acropolis'. In P. Carabott, Y. Hamilakis, and E. Papargyriou (eds), *Camera Graeca: Photographs, Narratives, Materialities*, London: Ashgate, pp. 133–157.

Harmanşah, Ö. 2015. ISIS, Heritage, and the Spectacles of Destruction in the Global Media. *Near Eastern Archaeology* 78: 170–177.

Helbig, D.K. 2016. *La trace de Rome?* Aerial Photography and Archaeology in Mandate Syria and Lebanon. *History of Photography* 40: 283–300.

Hoffman, G. and L. Brody 2011. *Dura-Europos: Crossroads of Antiquity*. Boston: McMullen Museum of Art.

James, S. 2004. *The Excavations at Dura Europos 1928–1937. Final Report VII, The Arms and Armour and other Military Equipment*. London: British Museum Press.

Latour, B. 2002. What is Iconoclash? Or Is There a World beyond the Image Wars? In P. Weibel and B. Latour (eds) *Iconoclash, Beyond the Image-Wars in Science, Religion and Art* (edited by Peter Weibel and Bruno Latour). Karlsruhe: ZKM, pp. 14–37.

Lucas, G. 2001. *Critical Approaches to Fieldwork. Contemporary and Historical Archaeological Practice*. London: Routledge.

Lucas, G. 2005. *The Archaeology of Time*. London: Routledge.

Lucas, G. 2012. *Understanding the Archaeological Record*. Cambridge: Cambridge University Press.

McFadyen, L., H. Lewis, N. Challands, A. Challands, D. Garrow, S. Poole, M. Knight, N. Dodwell, D. Mackay, L. Denny, P. Whitaker, P. Breach, L. Lloyd-Smith, D. Gibson and P. White 1997. Gossiping on Other People's Bodies. Unpublished paper presented at Theoretical Archaeology Group (TAG) 1997 Conference, Bournemouth University.

Mairs, R. and M. Muratov 2015. *Archaeologists, Tourists, Interpreters: Exploring Egypt and the Near East in the Late 19th–Early 20th Centuries*. London: Bloomsbury.

Morgan, C. 2016. Analog to Digital: Transitions in Theory and Practice in Archaeological Photography at Çatalhöyük. *Internet Archaeology*. http://intarch.ac.uk/journal/issue42/7/index.html

Mulvey, L. 2000. The Index and the Uncanny. In C.B. Gill (ed.) *Time and the Image*. Manchester: Manchester University Press, pp. 139–148.

Murray, T. 1999. A Return to the 'Pompeii Premise'. In T. Murray (ed.) *Time and Archaeology*. London: Routledge (One World Archaeology 37), pp. 4–27.

Neep, D. 2012. *Occupying Syria under the French Mandate: Insurgency, Space and State Formation*. Cambridge: Cambridge University Press.

Olivier, L.C. 2001. Duration, Memory, and the Nature of the Archaeological Record. In H. Karlsson (ed.) *It's about Time: The Concept of Time in Archaeology*. Gothenburg: Bricoleur Press, pp. 61–70.

Olivier, L. 2011. *The Dark Abyss of Time: Archaeology and Memory*. Lanham, MD: AltaMira Press.

Ousterhout, R.G. 2011. *John Henry Haynes. A Photographer and Archaeologist in the Ottoman Empire 1881–1900*. London: Cornucopia.

Papalexandrou, A. and M. Mauzy 2003. The Photographs of Alison Frantz. Revealing Antiquity Through the Lens. *History of Photography* 27: 130–143.

Pillet, M. 1928. *Rapport sur les travain de Doura-Europos*. Avril–Mai 1928. Unpublished report in Dura Archive, Yale University Art Gallery.

Poidebard, A. 1934. *La Trace de Rome dans le désert de Syrie. Le Limes de Trajan a la conquête arabe. Recherches aériennes (1925–1932)*. Paris: Libraire orientaliste Paul Geuthner.

Quirke, S. 2010. *Hidden Hands. Egyptian workforces in Petrie excavation archives, 1880–1924*. London: Duckworth.

Rostovtzeff, M.I., A.R. Bellinger, F.E. Brown and C.B. Welles (eds) 1944. *The Excavations at Dura-Europos conducted by Yale University and the French Academy of Inscriptions and Letters. Preliminary Report on the Ninth Season of Work, 1935–1936. Part 1, The Agora and Bazaar*. New Haven, CT: Yale University Press.

Shanks, M. and C. Tilley 1987. *Social Theory and Archaeology*. Albuquerque: University of New Mexico Press.

Shepherd, N. 2003. 'When the Hand that holds the Trowel is Black . . .': Disciplinary Practices of Self-Representation and the Issue of 'Native' Labour in Archaeology. *Journal of Social Archaeology* 3: 334–352.

Talbot, W.H.F. 1844. *The Pencil of Nature*. London: Longman, Brown, Green and Longmans.

Zacharia, K. 2015. Nelly's Iconography of Greece. In P. Carabott, Y. Hamilakis and E. Papargyriou (eds) *Camera Graeca: Photographs, Narratives, Materialities*. London: Ashgate, pp. 233–256.

5
PARAFICTIONS: A POLAROID ARCHAEOLOGY

Joana Alves-Ferreira

Figure 5.1 'Atlas: photographing photographs' – Original Polaroid taken at the archaeological site of Castanheiro do Vento (Vila Nova de Foz Côa, Portugal), 2009 (photo: Joana Alves-Ferreira).

Figure 5.2 'Diffused Reality: Space, memory, text' – Original Polaroid taken at the archaeological site of Castanheiro do Vento (Vila Nova de Foz Côa, Portugal), 2009 (photo: Joana Alves-Ferreira).

Figure 5.3 'To read what was never written' – Original Polaroid taken at the archaeological site of Castanheiro do Vento (Vila Nova de Foz Côa, Portugal), 2009 (photo: Joana Alves-Ferreira).

Figure 5.4 'At a moment of danger' – Montage exercise on original Polaroids taken at the archaeological site of Castanheiro do Vento (Vila Nova de Foz Côa, Portugal), 2009/2015 (photo: Joana Alves-Ferreira).

Figure 5.5 `Where is the Nymph?' – Montage exercise on original Polaroids taken at the archaeological site of Castanheiro do Vento (Vila Nova de Foz Côa, Portugal), 2009/2015 (photo: Joana Alves-Ferreira).

*

This chapter undertakes an archaeology of the Polaroid camera in order to explore the connections between archaeological photography and time. It introduces the idea of 'parafiction' to complement conventional notions of objectivity and authenticity on the one hand, and representation and the construction of archaeological knowledge on the other, in the study of archaeological photography.

The Polaroid camera and its self-developing film bring image and object together in time and space, iconographically and aesthetically. In the context of Archaeology, this 'instant photography' reveals certain things about photography and time (see Figure 5.1). The idea of 'instant' archaeological photography evokes a similar paradox to that which accompanies the notion of 'contemporary archaeology' – the prospect of an archaeology of our own moment in time. By describing the Polaroid as instant photography we rely on a particular attitude towards the measurement of time – an idea consolidated in Western thought only in the last decades of the nineteenth century. This is the idea of the absolute instant or the modern experience of the present moment (cf. Knight and McFadyen this volume).

As with contemporary archaeology, so with instant photography: let us imagine it not so much as a simultaneity as an intervention in time. As an experience, does the sudden moment of the Polaroid mean that any process of verification must be inherently improvised and partial? How much of what emerges in the moment between the subject and the exterior world is a materialization of already imagined images? In that temporal gap, how much does objective recording make room for creativity?

If the Polaroid picture, as 'instant photography', operates at the boundary between fact and fiction, then perhaps an 'interchangeability' of images through history, memory and storytelling occurs for contemporary archaeology too. If so, archaeological photography may represent a unique space in which to confront conventional conceptions of how history is written and then crystalized, questioning the very notion of the 'original document' which always requires a retrospective kind of reading – and opening up the possibilities of a speculative visual archaeology.

*

The first thing I found was this. What the Photograph reproduces to infinity has occurred only once: the Photograph mechanically repeats what could never be repeated existentially. In the Photograph, the event is never transcended for the sake of something else: the Photograph always leads the corpus I need back to the body I see; it is the absolute Particular, the sovereign

Contingency, matte and somehow stupid, the *This* (this Photograph, and not Photography), in short, what Lacan calls the *Tuché*, the Occasion, the Encounter, the Real, in its indefatigable expression . . . Whatever it grants to vision and whatever its manner, a photograph is always invisible: it is not it that we see.

BARTHES 1981 [1980]: 4; 6

A photograph collapses spatial time and human time within a single frame. In doing so it appears to remove any subjectivity from the image (Barthes (1981 [1980]: 21), bringing the bodily experience of space into interior consciousness, as if the camera were leaning forward to look only within the spot of light. The collapse is one of subjectivity and objectivity, fiction and science. 'Any purely reflexive discourse runs the risk of leading the experience of the outside back to the dimension of interiority', warns Foucault, while

The vocabulary of fiction is equally perilous: due to the thickness of its images, sometimes merely by virtue of the transparency of the most neutral or hastiest figures, it risks setting down ready-made meanings that stitch the old fabric of interiority back together in the form of an imagined outside.

FOUCAULT 1994 [1966]: 523[1]

As a nineteenth-century technology, Photography (etymologically, 'writing with light') seeks to meet the aspirations of a certain kind of scientific vision that might penetrate the opacities of the body and of the world, a belief that knowledge of both interior and exterior can be collapsed into the image as an optical form of knowledge. To seek to understand photography as knowledge, as both expanded observation and contracted experience, politically situated through their construction of neutrality, is to attend to images, devices and mechanisms involved in the technical production of a 'collective view, a visual culture' (Sicard 1998: 11).[2] A central element of the photograph as a device is that it can translate images into concepts and concepts into images, producing a sense of the photograph as an irrevocable document of presence. As Sontag puts it,

Photographs are, of course, artifacts. But their appeal is that they also seem, in a world littered with photographic relics, to have the status of found objects – unpremeditated slices of the world.

SONTAG 1977: 68–9

All photographs promote nostalgia, perhaps (Sontag 1977: 15), but the taking of a Polaroid picture offers a special example of such revelations of the authentic through its distinctive aura of processed nostalgia – its particular pathos, a sort of magical and sentimental dimension that comes to light in the form of a neat 'slice' of time, somehow taking place for a second time before our eyes. As an event, Polaroid not only slices the world but creates a kind of mythical illusion of sudden antiquity of the moment that just passed. As a tool and a medium, we can use the Polaroid to interrogate the photographic object and its relationship with memory: its double iconography, of both the subject of the image and of the object of the photograph as an artefact, between which there are what Benjamin (1977 [1931]) described as the 'tiny sparks of contingency, of the here and now' (Figure 5.2).

*

From the late nineteenth century, the modern experience of time has involved a growing acceleration – the progressive shortening of the distances between one moment and another and in the mechanization of mass productive processes, within which the Polaroid fits as a form of instant memory. (Archaeology has been implicated in this process too.) But modern experience has also involved an increasing resistance to such acceleration, in the form of suspended orientations towards the future as source of values and a guide for action (Lissovsky 2008). The increasing speeds of retention and memory have thus been accompanied by anticipations that, for the photographic image, involve not just presence but the absent future – what is not yet visible (cf. Barrento 2005: 43). How to expand the Benjaminian use of Photography, of the flash of the photographic image, to corroborate a micrological conception of the historical event as a particular, legible moment in time – history actualized as time filled by the presence of the now (*Jetztzeit*; see Benjamin 1999 [1940]: 252–3) – to account for this future orientation? Perhaps by complementing Benjamin's account of the illumination or awakening of history through a conception, borrowing from Didi-Huberman, of the discovery of the past as kind of 'breach' in history.[3] As an event, the Polaroid resists the instantaneous, breaches the process of an ever-accelerated contemporaneity (see Figure 5.3).

*

Now you press the red electric button . . . and there it is. You watch your picture come to life, growing more vivid, more detailed, until minutes later you have a print as real as life. Soon you're taking rapid-fire shots as you search for new angles or make copies on the spot. The SX-70 becomes like a part of you, as it slips through life effortlessly . . .

POLAROID'S SX-70 advertisement, 1975

The Polaroid was marketed as an 'instant document'. But as an icon of the modern experience of time, there is more to it than ultimate synchrony. It transformed the photographer from an offstage observer to a part of the photographic event in a new way, bringing a new texture to immediate memory, bringing the world closer (Bonanos 2012: 104). As well as the modern, ocular, chronometrical ideal of reconciling the instant with duration (Thomas 2008), the *dynamics* of the Polaroid as an event also expanded, as if illuminated from within by the certainty of its own immediate future, to encompass something of what is yet to come.[4] Film-maker André Tarkovsky has expressed something like this future-orientation of the image as a breach, in terms of life and hope:

> An artistic image is one that ensures its own development, its historical viability. An image is a grain, a self-evolving retroactive organism. It's a symbol of actual life, as opposed to life itself. Life contains death. An image of life, by contrast, excludes it, or else sees in it a unique potential for the affirmation of life. Whatever it expresses – even destruction and ruin – the artistic image is by definition an embodiment of hope.
> TARKOVSKY 1994, cited in CHIARAMONTE and TARKOVSKY 2014: 56

So, the Polaroid is as much a gesture into the near future, an enunciative proposition of resistance to the past catching up with the present moment, as it is an instantaneous document. As a device for the 'writing of resistance' in the medium of light, it is like a reservoir for the event rather than just its simulacrum, in that it holds within it possibility, the instant as chance, as a 'tameless irruption' (Vilela 2010: 524–5). Where writing with a pen or a keyboard allows time to think before the words are formed, the Polaroid pushes to the limit the process through which the fleeting nature of the photographic defers anguished thought to after the ever-disappearing moment, after the 'decisive instant'.[5]

Here, absences foreshadow presence as well as mourning loss. The photographic subject must always pass us by, a trace of what no longer is, but Polaroid reveals how the photograph in the same moment bears witness to an event that is unceasing, ongoing, precarious rather than just veracious, silent as well as meaningful, in shadow as well as lit up in a Benjaminian flash, hopeful as well as retrospective, nomadic as well as static (Cf. Prado Coelho 1988: 679), testimony of what is to come as well as evidence of what has already occurred. Photography as, to use Tarkovsky's terms, images of life as well of death. Or, to borrow from Martí Peran, as images of virtuality, porosity and possibility not just of an 'empty leftover':

> Of the real we know no more than its trace, a presence that cannot designate more than an absence; a testimony of what no longer is. The simile of the trace, so recurrent in contemporary thought and, for obvious reasons, extremely

close to the specificity of photographic language, has an added virtue, however. As an empty leftover of the real, the trace accentuates the virtuality of the image that remains. Despite having – or precisely because of having – a specific size and texture that lend it a quality of reality, the trace contains nothing, it doesn't store meaning, but a recollection of the absent, hence the rapidity with which it is imposed in a strictly physical manner, as a mere porous thing that no longer belongs to the world of representation, but to that of possibility.

PERAN 2000: n.p.

The Polaroid resists the instantaneous precisely because of its condition of being as an artefact. Its hyper-immediacy collapses any idea of the permanence of the fixedness of a past moment in time into its obverse: Photography as an event that does not stop, the visual manifestation of the passage of time, a trace that is somehow unprecedented, transforming.

Even the Polaroid ages. It grows old, as if it were alive. Its temporal surface is porous. The changeless frame of a frozen instant gives way to constantly changing repetition of fluid and tenuous movements from impression to hesitation and expectation. This subtle beyondness of the temporal image holds much in common with that Barthes called 'the *kairos* of desire' (Barthes 1981 [1980]: 59). As the physical evidence of something that never ceases to happen, the Polaroid, with its distinctive layout and texture, is a unique body for the potential writing of that what only belongs to possibility.

*

If the Polaroid shows us something more generally true about Archaeology and Photography, it is that the photograph constitutes not only a trace of the past, but also a speculation about the future. How does this shape our conception of archaeological knowledge? In his book *Experiencing the Past*, Michael Shanks introduced the idea of an 'archaeological poetics'

Invention; non-identity and the necessity of going beyond what I have found; being drawn into metaphor and allegory. As an archaeologist, what constructions might I make? If the facts slip away so easily, how might I represent the past? These are the concerns of an archaeological poetics.

SHANKS 1992: 142

But can we push his call for 'experimentation' beyond idea of construction and representation – beyond, that is, reducing fact to fiction. Push harder at the boundary between fact and fiction in Archaeology as documentation?

I want to suggest that the idea of 'parafiction' might be of use here. As a prefix, *para* 'suggests that which lies "alongside of", "by", "past", or "beyond"'

(Rother 1976: 34). As a 'post-simulacral strategy', parafiction may be less about the construction of facts, or rendering truth through fiction, and 'oriented less toward the disappearance of the real than toward the pragmatics of trust' (Lambert-Beatty 2009: 54). As with the documentary film-making of Salomé Lamas, it can 'investigate the traumatically repressed, the seemingly un-representable, and the historically invisible' through new combinations of the fictional and the documentary (Lamas 2016). We might map these combinations onto the documented past and the speculative future, imagining the parafictional as a form of standpoint:

> To know it is necessary to take a position [. . .]. To take a position is to be situated at least twice, on at least two fronts at least that comprises any position since any position is, fatally relative. For example, it's about facing something [. . .] It's also about situating yourself in time. To take a position is to desire, to demand something, to situate oneself in the present and aim for a future.
>
> DIDI-HUBERMAN 2009: 11[6]

Archaeological photography, like the Polaroid, documents the future as much as it imagines the past. Far from representing the instantaneous or the flashback it renders the present moment as something unfixed, fluctuating and uncertain (cf. Deleuze 1985; Didi-Huberman 2011) (Figure 5.4).

As a parafiction, it is a form that thinks; a territory where we can imagine another possible geography, formed of chance, contingency and possibility; where we can, as Benjamin had it, 'read what was never written' (Alves-Ferreira 2017). A reserve for visual imagination, and thus for temporal knowledge (Figure 5.5).

Acknowledgements

I would like to thank the *Foundation for Science and Technology* (FCT – Portugal) for funding my research and the *Centro de Estudos em Arqueologia, Artes e Ciências do Património* (CEAACP) for their support. In particular, I would like to thank Ana Vale, Sérgio Gomes, and Andrew May for the comments and corrections on previous versions of this text. To my supervisor, Susana Soares Lopes, for her support and guidance. To my brother, Luís Laugga, for sharing the passion for Polaroid and for his always inspiring reflection. Finally, my deep gratitude goes to Lesley McFadyen and Dan Hicks for their kind and challenging invitation to contribute to this volume.

Notes

1 'Expérience qui devait rester alors non pas exactement enfouie, car elle n'avait pas
   pénétré dans l'épaisseur de notre culture, mais flottante, étrangère, comme
   extérieure à notre intériorité, pendant tout le temps où s'est formulée, de la façon la
   plus impérieuse, l'exigence d'intérioriser le monde, d'effacer les aliénations, de
   surmonter le moment fallacieux de l'Entäusserung, d'humaniser la nature, de
   naturaliser l'homme et de récupérer sur la terre les trésors qui avaient été dépensés
   aux cieux [. . .]. Tout discours purement réflexif risque en effet de reconduire
   l'expérience du dehors à la dimension de l'intériorité ; invinciblement, la réflexion tend
   à la rapatrier du côté de la conscience et de la développer dans une description du
   vécu où le 'dehors' serait esquisser comme expérience du corps, de l'espace, des
   limites du vouloir, de la présence ineffaçable d'autrui. Le vocabulaire de la fiction est
   tout aussi périlleux: dans l'épaisseur des images, quelquefois dans la seule
   transparence des figures des plus neutres ou les plus hâtives, il risque de déposer
   des significations toutes faites, qui, sous les espèces d'un dehors imaginé, tissent à
   nouveau la vieille trame de l'intériorité'.

2 'Il n'est pas dans l'objet de cet ouvrage de décerner des bons points ; de trier ce
   qui, de l'oiseau ou du poisson, fut bien observé ou mal reproduit. Mais de
   comprendre comment se fabrique un regard collectif, une culture visuelle: par quels
   effets, sous l'emprise de quelles images, de quels appareils, à l'aide de quels
   mécanismes de légitimation. Car les industries du savoir s'enchevêtrent intimement
   avec celles du croire et leur corollaire: celles du faire croire. Plus s'affirme – au
   premier étage – la méconnaissance des dispositifs de vision, mieux s'exerce – au
   deuxième – la fonction politique des images. C'est en affichant leur neutralité qu'elles
   transmettent le mieux des points de vue délibérés; en installant des faits qu'elles
   fonctionnent comme fictions. En clamant leur indépendance qu'elles soudent et
   parlent culture. Documents et enchantements: les images savantes réussissent ce
   tour de passe-passe de certifier et d'émouvoir à la fois.'

3 As Didi-Huberman writes in *Images Malgré Tout*, 'each discovery *emerges from it
   like a breach in the history conceived*, a provisionally indescribable singularity that
   the researcher will attempt to weave into the fabric of everything he or she already
   knows, in order to produce, a *rethought history* of the event in question' (Didi-
   Huberman 2004; author's translation).

4 'Loin d'appeler l'intériorité à se rapprocher d'une autre, l'attirance manifeste
   impérieusement que le dehors est là, ouvert, sans intimité, sans protection ne
   retenue (comment pourrait-il en avoir, lui qui n'a pas d'intériorité, mais se déploie à
   l'infini hors de toute fermeture ?) [. . .]; il ne peut pas s'offrir comme une présence
   positive – chose illuminé de l'intérieure par la certitude de sa propre existence –,
   mais seulement comme l'absence que se retire au plus loin d'elle-même et se
   creuse dans se signe qu'elle fait pour qu'on avance vers elle, comme s'il était
   possible de la rejoindre' (Foucault 1994 [1966]: 526).

5 Cartier-Bresson's preface to *Images à la sauvette* (1952), titled 'L'Instant décisif',
   reflects how: 'de tous les moyens d'expression, la photographie est le seul que fixe
   un instant précis. Nous jouons avec des choses qui disparaissent et quand elles ont
   disparu, il est impossible de les faire revivre. On ne retouche pas son sujet ; on peut
   tout au plus choisir parmi les images recueillies pour la présentation du reportage.
   L'écrivain a le temps de réfléchir avant que le mot ne se forme, avant de le coucher

sur le papier ; il peut lier plusieurs éléments [. . .] Pour nous, ce qui disparaît, disparaît à jamais de là, notre angoisse et aussi l'originalité essentielle de notre métier.' (Cartier-Bresson 1952: 19).

6 'Pour savoir il faut prendre position [. . .]. Prendre position, c'est se situer deux fois au moins, sur les deus fronts au moins que comporte toute position puisque toute position est, fatalement relative. Il s'agit par exemple d'affronter quelque chose [. . .] Il s'agit également de se situer dans le temps. Prendre position, c'est désirer, c'est exiger quelque chose, c'est se situer dans le présent et viser un futur.'

References

Alves-Ferreira, J. 2017. The Art of Endangering Bodies: A First Movement on 'How to Read What Was Never Written'. In A. Vale, J. Alves-Ferreira and I. Garcia Rovira (eds) *Rethinking Comparison in Archaeology*. Newcastle: Cambridge Scholars, pp. 13–39.

Barrento, J. 2005. *Ler o que não foi escrito. Conversa inacabada entre Walter Benjamin e Paul Celan*. Lisboa: Edições Cotovia.

Barthes, R. 1981 [1980]. *Camera Lucida. Reflections on Photography* (trans. R. Howard). New York: Farrar, Straus & Giroux.

Benjamin, W. 1977 [1931]. Kleine Geschichte der Photographie. In *Gesammelte Schriften Volume 2.* Frankfurt: Suhrkamp, pp. 368–385.

Benjamin, W. 1999 [1940]. Theses on the Philosophy of History. In H. Arendt (ed.) *Illuminations* (trans. H. Zorn). London: Pimlico, pp. 245–255.

Bonanos, C. 2012. *Instant. The Story of the Polaroid*. New York: Princeton Architectural Press.

Cartier-Bresson, H. 1952. *Images à la sauvette*. Paris: Verve.

Chiaramonte, G. and A. Tarkovsky (ed.) 2014. *Instant Light. Tarkovsky Polaroids*. London: Thames and Hudson.

Deleuze, G. 1985. *L'Image-Temps – Cinéma 2*. Paris: Éditions de Minuit.

Didi-Huberman, G. 2004. *Images malgré tout*. Paris: Éditions de Minuit.

Didi-Huberman, G. 2009. *Quand les images prennent position. (L'œil de l'histoire, I)*. Paris: Éditions de Minuit.

Didi-Huberman, G. 2011. Opening the Camps, Closing the Eyes: Image, History, Readability. In G. Pollock and M. Silverman (eds) *Concentrationary Cinema: Aesthetics as Political Resistance in Alain Resnais's 'Night and Fog' (1955)*. New York: Berghahn, pp. 84–125.

Foucault, M. 1994 [1966]. La pensée du dehors. In *Dits et Écrites I (1954–1969)* (eds, D. Defert and F. Ewald). Paris: Gallimard, pp. 518–539.

Lamas, S. 2016. *Parafiction: Selected Works*. Milan: Mousse Publishing.

Lambert-Beatty, C. 2009. Make-Believe: Parafiction and Plausibility. *October* 129: 51–84.

Lissovsky, M. 2008. The Photographic Device as a Waiting Machine. *Image [&] Narrative* 23. http://www.imageandnarrative.be/inarchive/Timeandphotography/lissovsky.html

Peran, M. 2000. Reality, Porous and Besieged. In J. Fontcuberta, J. Wagensberg and M. Peran (eds) *Joan Fontcuberta – Twilight Zones*. Barcelona: Actar (unpaginated).

Prado Coelho, E. 1988. *A Noite do Mundo*. Lisboa, Imprensa Nacional Casa da Moeda.

Rother, J. 1976. Parafiction: The Adjacent Universe of Barth, Barthelme, Pynchon, and Nabokov. *boundary 2* 5(1): 20–44.

Shanks, M. 1992. *Experiencing the Past*. London: Routledge.

Sicard, M. 1998. *La Fabrique du Regard. Images de Science et Appareils de Vision (XVᵉ –XXᵉ siècle)*. Paris: Éditions Odile Jacob.

Sontag, S. 1977. *On Photography*. London: Penguin.

Tarkovsky, A. 1994. *Time Within Time: the Diaries 1970–1986* (trans. K. Hunter-Blair). London: Faber and Faber.

Thomas, J. 2008. On the Ocularcentrism of Archaeology" In J. Thomas and V. Oliveira Jorge (eds) *Archaeology and the Politics of Vision in a Post-Modern Context*. Newcastle: Cambridge Scholars, 2008), pp. 1–12.

Vilela, E. 2010. *Silêncios Tangíveis. Corpo, Resistência e Testemunho nos Espaços Contemporâneos de Abandono*. Porto: Edições Afrontamento.

6

ARCHAEOLOGY, PHOTOGRAPHY AND POETICS

Sérgio Gomes

Here are four photographs from the archive of an excavation that I directed in 2011 at Porto de Moura 2, a prehistoric site located in Alentejo in Portugal (Figures 6.1–6.4). I can recall the circumstances of our taking of these photographs very well. The work was undertaken as a part of commercial archaeology in advance of the construction of a pipeline. The project ran during the late summer and the beginning of the autumn, the last days of the hot weather and the first of the harvest rain (Figure 6.1). One day we were faced with quite dry sediments that sometimes seemed like concrete, and the next we were excavating soft, muddy sediments that needed cleaning all the time. More and more often we were faced with flooded trenches that had to be drained. The path to the site got muddier so the van often got stuck and we had to get out to push it. After getting the van to the site, there was bailing out water in the trenches, and the general site clean-up before we could start excavating again, take the camera out of the van and document the people, trowels, buckets, bags, labels, pencils, sediments, rocks and artefacts – and the forces that run between them (Figures 6.2 and 6.3).

There is an intimacy between Archaeology and Photography, as many have previously observed (e.g. Bateman 2005, 2006; Bohrer 2005; Moser 2012; Shanks 1992, 1997; Shanks and Svabo 2013; Thomas 2008) (Figure 6.4).

* The writing of this chapter was supported by the FCT (Fundação para a Ciência a Tecnologia, Portugal) under the grant SFRH/BPD/100203/2014. I would like to thank Diogo Cão, Susana Lopes, Ana Vale and Joana Alves for their help with the ideas in this paper. My deepest gratitude to all the colleagues with whom I had the privilege to work at Porto de Moura 2. I also thank Lesley McFadyen, Dan Hicks and Julia Roberts for editing my text.

Figure 6.1 Archaeological excavations at Porto de Moura 2, Alentejo, Portugal, 2011 (photo: Sérgio Gomes).

Figure 6.2 A working shot taken during archaeological excavations at Porto de Moura 2, Alentejo, Portugal, 2011 (photo: Sérgio Gomes).

Figure 6.3 A working shot taken during archaeological excavations at Porto de Moura 2, Alentejo, Portugal, 2011 (photo: Sérgio Gomes).

Figure 6.4 A working shot taken during archaeological excavations at Porto de Moura 2, Alentejo, Portugal, 2011 (photo: Sérgio Gomes).

Archaeology and Photography come together through a series of smaller relationships. These are made between a variety of shifting elements: the camera and light, the landscape past and present, the archaeologist as excavator and as photographer, and so on. These different actors are involved even when they cannot be seen in the image. The archaeological photograph involves a distinctive play of forces between what is being portrayed and what is holding the frame. It is a play of presence and absence, forces that are negotiated through the conventional codes, experiences and objectives of Archaeology as a discipline. The camera thus, in the hands of the archaeologist, becomes an apparatus that changes how we make knowledge of the past.

Archaeological photographs convey information in visual and material form about an archaeological reality: across space through disciplinary circulations, and across time in the archive and the museum (cf. Molyneaux 1997; Smiles and Moser 2005). They create mediascapes that circulate knowledge of the past, and that constitute new spaces for dialogue or exchange. Many have observed how a photograph can connect Archaeology with other kinds of media or social networks, so bringing about new relationships far beyond those intended or imagined by the archaeologist (e.g. Clack and Britain 2007; Holtorf 2007; Thomas and Jorge 2008). But there can equally be far less than intended or imagined present in the image. At Porto de Moura 2 the camera was used to create quiet and peaceful photographs – the forces that made the scene possible, whether seen or unseen, are silenced and ineffable (Figure 6.1). As I worked through the clean images in the photographic archive, none of the messy and muddy experiences I recalled were recorded there.

*

In *The Life of Infamous Men* (1979) Foucault discusses the emergence of *lettres de cachet*, a bureaucratic tool used in France during the seventeenth and eighteenth centuries, in which the king's subjects could report directly to the king about their everyday problems. The emergence of this material support facilitated the redistribution of information that came from diverse discourse practices, the collection of which made possible new discourse practices (with new agents and new possibilities in which to act). The writing of *lettres de cachet* created the conditions of communication between the king and his subjects – a textual space in which direct communication could take place. Everyday life was mediated and put into words. Their collection facilitated the emergence of a knowledge of daily life that was then incorporated into state affairs. This in turn became a new set of juridical forms.

We could draw a comparison between *lettres de cachet* and the written documents, like filled-out context sheets, that are made through archaeological recording in the field. Both act as frameworks that allow participants to put experiences into words and then make this information sharable with others. We might extend our analogy to the written labels used to record artefacts: labels ask about the nature of material, about its location and its context, acting as a passport for past fragments in which there is a description of how they are to be managed in archaeological affairs. As *lettres de cachet* are about framing life in words, so the archaeologist's written context sheets and labels are about framing the archaeological process (Figure 6.2).

The analogy might even stretch to encompass my 'quiet and peaceful' archaeological photographs. They too code, frame, translate and communicate experience, but through images rather than words. Both the words and the photographs of archaeological fieldwork create spaces of documentation. Like the textual spaces created by Foucault's *lettres*, they reshape reality, and thus redistribute power. Any firm opposition between documentation and objectivity is in vain. Lorraine Daston and Peter Galison (1992, 2007), in their classic discussion of changing conceptions and practices of objectivity, show the role that different ways of making and using images has played in the constitution of scientific knowledge in different disciplinary projects. Particular modes of objectivity, from 'truth-to-nature' to 'mechanical objectivity' to more recent forms of reflexivity, are never separate from the perhaps limitless variety of scientists' theories of their own personhood and modes of detachment or implication. Thus, the changing role of objectivity in the production of knowledge is not a history of different weightings or emphases along a continuum from objectivity and subjectivity: it is a history of triangulation from these two imagined poles and the constitution of the scientific self (Figure 6.3).

*

Archaeology re-configures and transforms the world. This includes Photography as much as excavation. As Michel de Certeau (1988: 75) writes, 'a work is "scientific" when it produces a *redistribution* of space'. Michael Shanks points to collage, montage and quotation as key elements of archaeological photography, and we might highlight these as modes of intervening and transforming or 'redistributing' the world (Shanks 1992: 149–51). Shanks introduces the idea of an 'archaeological poetics' as 'strategies for representing the dynamic object past' and 'aspects of archaeology as craft', to work through a series of tensions: 'subjective and objective, particular and general, fragments and construction, experiment and responsibility, pluralism and authority' (Shanks 1992: 168). How can this idea be pushed to put the visual and the textual into a new dialogue?

During fieldwork and post-excavation, the act of taking a photograph is performed many times, under both formal and informal 'working shot' conditions. The more irregular images, for example those that more informally show people, whether excavators or others, in the frame rather than confirming strictly to formal methods of measuring rod and cleaned surface, represent unique documents of the practices through which these redistributions take place (Figure 6.4). The archaeological process is a material mediation focused on the translation of the indexability of something that we aim to know into the indexability of a disciplinary framework allowing us to be archaeologists. As Gavin Lucas (2012: 228–31) emphasizes, within this process of mediation and intervention past materials are not just given meaning but are materialized. Material translation, as Lucas (2012: 238–9) puts it, is not a game of equivalences between an original and a copy, but is developed through the overcoming of the asymmetry entailed in the opposition between an idea of the original and the copy. Translating, as mediation and redistribution, is a matter of transformation within which we shape our study objects and create the conditions and the projects to know them. Translation in Archaeology is a material dialogue between the particularity of materials and the demands of the disciplinary project, the transformation of something into something else. This metamorphosis of past materials is screened through residuality (Lucas 2012: 204–14) and material memory (Olivier 2012: 187–94). Archaeological photography, in the context of these temporal changes in materials, documents by translating material flows, translating the *ineffable* in a transformation (Shanks 1992: 102). Photographs engage with how we dwell on the silence of past materials, and make present how we engage with the silence of other agents within the archaeological intervention; how silent but vibrant matter (Bennett 2010) is transformed through fieldwork, translating between silence and visibility. Photography can reveal what has been excluded in a manner analogous to discovery through excavation. In this sense, archaeological photographs do not so much document a moment in time as create new geographies through the redistribution of what is seen and what is unseen (Figure 6.3).

*

Archaeological writing and photography redistributes, transforms and translates past materials. In comparing words and images in this manner, we can borrow from João Barrento's account of poetry (2014). Barrento (2014: 11–16) outlines different modes through which poetry engages with the world: responding to emotion, resisting hegemony, but also making a record of the world. For Barrento, in none of these cases is a poem just a way to interpret or describe the world; it is a future-oriented method for paying attention and thus creating the conditions for something new – 'enacting what is happening. . .poetry simply shows what there is: to see things, and show them, that is its modest utopia' (Barrento 2014: 19, my translation). Thus, in a discussion of the work of Maria Gabriela Llansol, Barrento shows how her writings are not a symbolic recording of the world, nor a prophecy of a utopic world, but a note on what is happening in a given moment at a given place. Through writing the world, Llansol is producing a memory that will act upon the conditions of the future. Writing is thus about transforming what is happening into an event; it is about taking it from *chronos* to *kairos* by comprehending its bindings and remaking its chain. By writing she understands how she is living and prepares the life to come. As she puts it, 'Poets see, and announce the immaterial geography to come' (Llansol 2000: 45, my translation). And like the operation of poetic text, as a way of seeing through description and thus as a means of transformation, so archaeological photographs enact the present not for the nostalgic preservation of past fragments, but to create the conditions for new knowledge. Words and images can thus represent sudden memories of the world that create the conditions for change and transformation (see Figure 6.4).

In Photography as in Archaeology, the absences of the past come to be known through the ongoing presences of past materials. De Certeau observes how:

> The figure of the past keeps its primary value of representing *what is lacking*. With a material which in order to be objective is necessarily *there*, but which connotes a past insofar as it refers first of all to an absence, this figure also introduces the rift of a future.
>
> DE CERTEAU 1988: 85

Photography and Archaeology, as intertwined crafts (Shanks 1992, 1997; Shanks and McGuire 1996; Bohrer 2005; Shanks and Svabo 2013), create the conditions for the announcement of new material geographies that contain immaterial pasts that are yet to come about.

My photographs – quiet and peaceful working scenes made possible after messy and muddy hands pushing a van – were a precondition for my memory of events that are, as I have described it here, invisible in them. For me to look at them now is to transform the experience and the knowledge I got during the excavation into something else, and to transform the images themselves, unpredictably. The photographs do not contain frozen actions but hold within them ongoing movements (Figure 6.1).

Writing is an art of describing material evidences with words. It translates the evidence into a text. In the syntax of the sentences and within the rhythm of their succession, there what is written prefigures what has not yet been read. So too in archaeological photography, which describes the past through the evidence of light, translating the experience of archaeological materials into the framed immaterial geography of the image.

When we interrogate the archaeological photograph we create a tension between what is seen and what is yet to be seen; as if it was a clue to a geography to come. By following that clue, we may start playing with a light that might show us places where we can try to take another glance, grasping a new direction in the materialization process.

References

Barrento, J. 2014. *Geografia Imaterial. Três Ensaios Sobre a Poesia*, Lisboa: Documenta.

Bateman, J. 2005. Wearing Juninho's Shirt: Record and Negotiation in Excavation Photographs. In S. Smiles and S. Moser (eds) *Envisioning the Past: Archaeology and the Image.* Oxford: Blackwell, pp. 192–203.

Bateman, J. 2006. Pictures, Ideas, and Things. The Production and Currency of Archaeological Images. In M. Edgeworth (ed.) *Ethnographies of Archaeological Practice: Cultural Encounters, Material Transformations*. Oxford: AltaMira Press, pp. 68–80.

Bennett, J. 2010. *Vibrant Matter: A Political Ecology of Things*. Durham, NC: Duke University Press.

Bohrer, F. 2005. Photography and Archaeology. The Image as Object. In S. Smiles and S. Moser (eds) *Envisioning the Past. Archaeology and the Image*. Oxford: Blackwell, pp. 180–191.

Clack, T. and M. Britain (eds) 2007. *Archaeology and the Media*. Walnut Creek, CA: Left Coast Press.

Daston, L. and P. Galison 1992. The Image of Objectivity. *Representations* 40: 81–128.

Daston, L. and P. Galison 2007. *Objectivity*. New York: Zone Books.

De Certeau, M. 1988. *The Writing of History* (trans. T. Conley). New York: Columbia University Press.

Foucault, M. 1979. The Life of Infamous Men (trans. P. Foss and M. Morris). In M. Morris and P. Patton (eds) *Michel Foucault: Power, Truth, Strategy*. Sydney: Feral Publications, pp. 76–91.

Holtorf, C. 2007. *Archaeology is a Brand! The Meaning of Archaeology in Contemporary Popular Culture*. Oxford: Archaeopress.

Llansol, M. G. 2000. *Onde vais Drama-Poesia?* Lisboa: Relógio D'Água.

Lucas, G. 2012. *Understanding the Archaeological Record*. New York: Cambridge University Press.

Molyneaux, B.L. (ed.) 1997. *The Cultural Life of Images. Visual Representations in Archaeology*. London: Routledge.

Moser, S. 2012. Archaeological Visualisation: Early Artefact Illustration and the Birth of the Archaeological Image. In I. Hodder (ed.) *Archaeological Theory Today*. Cambridge: Polity Press, pp. 292–322.

Olivier, L. 2012. *The Dark Abyss of Time. Archaeology and Memory* (trans. Arthur Greenspan). Lanham, MD: AltaMira Press.

Shanks, M. 1992. *Experiencing the Past: On the Character of Archaeology*, London: Routledge.

Shanks, M. 1997. Photography and Archaeology. In B.L. Molyneaux (ed.) *The Cultural Life of Images: Visual Representation in Archaeology*. London: Routledge, pp. 73–107.

Shanks, M. and R. McGuire 1996. The Craft of Archaeology. *American Antiquity* 61: 75–88.

Shanks, M. and C. Svabo 2013. Archaeology and Photography: A Pragmatology. In A. González Ruibal (ed.) *Reclaiming Archaeology: Beyond the Tropes of Modernity*. Abingdon: Routledge, pp. 89–102.

Smiles, S. and S. Moser (eds) 2005. *Envisioning the Past: Archaeology and the Image*. Oxford: Blackwell.

Thomas, J. 2008. On the Ocularcentrism of Archaeology. In J. Thomas and V.O. Jorge (eds) *Archaeology and the Politics of the Vision in a Post-Modern Context*. Cambridge: Cambridge Scholars, pp. 1–12.

Thomas, J. and V.O. Jorge (eds) 2008. *Archaeology and the Politics of the Vision in a Post-Modern Context*. Cambridge: Cambridge Scholars.

7

DURATION AND REPRESENTATION IN ARCHAEOLOGY AND PHOTOGRAPHY

Antonia Thomas

Late February 1925, Brodgar Farm, Stenness, Orkney

It had already been a long, hard winter. Like all Orkney farmers James Wishart was keen to get on with the business of getting his land ploughed, harrowed and sown for the next season's silage crop. But his field contained rather a lot of sizeable and awkward stones, which would have to be removed before they damaged his plough. He had pulled out a good number when a large flagstone slab caught his eye: strikingly carved along one of its edges, with a pattern of bands not unlike a Fair Isle sweater.

Wishart had to dig around the stone to remove it, and more objects began to emerge: two balls of stone, just bigger than his fist and smooth like beach pebbles. But the field needed ploughing, and it wouldn't clear itself of these rocks. He placed his finds by the dyke and carried on. He was little interested in old stones, carved or not, and there was work to be done. Even so Wishart thought he would mention the matter to his neighbour Peter Leith, who liked to look at such things. Leith had a keen knowledge of local history and archaeology. He also happened to be one of the few people in the West Mainland who owned a camera.

---

* Peter Leith's 1925 image of the Brodgar Stone provided the springboard for many of the ideas in this paper and sincerest thanks go to his son, Peter Leith junior, for donating the print and discussing it and his father's work. Thanks also to Alison Sheridan at the National Museum of Scotland for allowing me to access and photograph the Brodgar Stone and to Daniel Lee.

Figure 7.1 The Brodgar Stone, photographed by Peter Leith in 1925. Reproduced with the kind permission of Peter Leith (junior).

Leith came by straight away. As soon as he saw the stone, he knew he wanted to photograph it. He placed the slab on top of a roll of barbed wire in the farmyard, set up his tripod and plates, and looked through the viewfinder. When he had first examined the stone with Wishart, the deep grooves of the incised lines were obvious. If the light caught them just right then they were as clear as if they had been painted. But now through the camera lens the carvings were barely visible. Leith took a stick of chalk from his pocket and rubbed it along each incised mark, blowing away the excess to highlight the lines. When he developed his glass-plate negatives the next day, he was happy enough. The exposure of the sky wasn't quite right, but the stone was just as he wanted, and the chalk-filled lines were as clear as day (Figure 7.1).

THOMAS 2016, xv

Introduction

This chapter explores duration and representation in archaeology and photography. Its point of departure is the photographic print reproduced in Figure 7.1. The paper is aged and torn, the photograph slightly overexposed, and the print poorly aligned with the edge of the photographic paper. The image shows a large stone slab decorated with eight bands of incised markings, resting on an *ad hoc* plinth formed by the circular lid of a wooden barrel, on top of a roll of barbed wire. There is a roughly built drystone wall in the background, and the various pieces of wood, metal and sacking on the ground suggest a hastily constructed scene. The subject of the picture is a piece of flagstone that was quarried and decorated in the Neolithic, and unearthed by accident during early spring ploughing in February 1925 in Orkney, Scotland. Now known as the Brodgar Stone, it was captured on film by the camera of amateur local photographer Peter Leith on the day of its discovery.

As an artefact, and as an image, this can be interpreted from a range of different perspectives. My concern in this chapter, however, is not with the stone's original archaeological context (on which see Thomas 2016). Nor is it with an interpretation of the photographer's technical ability, or the aesthetics of the image itself. Instead, this chapter will explore how a study of this first photograph of the stone, and subsequent images produced of it, can be useful in understanding the relationship between archaeology, photography and time. In particular, I wish to examine the nature of *duration*, and the role it plays in the representation of the past. Crucial to the discussion is the way in which both archaeology and photography are frequently considered as ways of 'capturing' the past. This notion rests upon an understanding of temporality purely as *chronology*, and leaves little space for an understanding of the multi-durational character of the past.

This chapter presents an alternative view. It challenges the idea that photographs capture a fixed moment in time, and instead considers them as multi-durational artefacts. I will argue that this has implications for how we understand and represent the nature and reality of the past in archaeology. Thinking through photography can thus provide an analogy for how we encounter, and work with, the visual and material culture of archaeology in the present.

Archaeology and photography

Over the past two decades, the shared disciplinary histories of archaeology and photography have been the subject of much discussion (e.g. Bohrer 2005; Bohrer 2011a; Cochrane 2018; Hamilakis and Ifantidis 2015; Shanks 1997; Shanks and Svabo 2013). As techniques and technologies, they both represent, as Michael Shanks and Connie Svabo have noted, 'constituting moments of modernity' (Shanks and Svabo 2013: 90). Archaeology and photography emerged in parallel during the mid-nineteenth century, as reciprocal endeavours that shared a common desire to apprehend past events. The objectifying process of photography lifts things out of the past, to be viewed in the present, 'like flies in amber' (Wollen

1984: 118). Archaeology operates in an analogous manner, through what Yannis Hamilakis, Aris Anagnostopoulos, and Fotis Ifantidis describe as 'the selective recovery, reconstitution and restoration of the fragmented material traces of the past' (Hamilakis *et al.* 2009: 285). Unsurprisingly, archaeological metaphors frequently appear in discussions of the photographic process (Bohrer 2011a: 8).

Archaeology and photography certainly share numerous interlinked concerns with the past and its representation. But to consider either archaeology or photography as ways of 'capturing' the past highlights a certain approach to time, and in particular duration, that requires further examination. It relies upon a notion of time purely as a uniform, linear and unidirectional *chronology* (Lucas 2005: 10). A chronological approach to time imagines it as made up of a series of instantaneous, fixed events akin to photographic *snapshots*. This bears little relation to the complexity of human experience, and effectively reduces the past to a series of 'still frames' (McFadyen 2013: 141). The past might be better understood as *durational*, comprising 'things that were produced and shaped in the past, but continue to live and exist in the present' (Hamilakis and Ifantidis 2015: 139; Olivier 2001: 61). This understanding might at first seem to run counter to a photographic understanding of time. But this durational quality, and the ability to enact multiple times simultaneously, is also common to photography (Hamilakis and Ifantidis 2015: 139). There is no such thing as an instantaneous photograph (Szarkowski 1980: 101), since even an apparent 'snapshot' contains a coalescence of times (cf. Plummer 2012: 36).

This is significant, as photography and archaeology also work together in a mutually constitutive role. Photography *mediates* the past, turning 'ancient sites and collections into textual and graphical forms that can be shared and discussed' (Shanks and Svabo 2013: 90). This brings with it a further entanglement. With its apparent ability to act as a 'time machine' (Badger 2007: 8), photography occupies a favoured role amongst modes of visual production and representation (Hunt and Schwarz 2010). Its apparent objectivity has meant that it has endured as the dominant medium of recording across a range of academic disciplines (Edwards 2011: 161) and photography has been at the heart of archaeological practice and representation since the outset (Shanks and Svabo 2013: 89). Theorizing photographs as multi-durational, however, disrupts any notion that photographs capture a fixed moment in time. It challenges assumptions that they might offer an immutable, objective representation of the past, and highlights the relationships between images and their representations, and between art and duration more generally (Baetens *et al.* 2010: xii). Exploring the multiple durations of photography can thus open up a parallel understanding of the temporal complexity that runs through archaeological material and practice. This allows us to examine the modes in which archaeologists have understood, represented and interpreted past visual culture – including artefacts such as the Brodgar Stone.

The Brodgar Stone: before and after

News of Wishart's 1925 discovery soon spread around local antiquarian circles. Later on that year a short note was published on the find by Dr Hugh Marwick in the *Proceedings of the Orkney Antiquarian Society*, a local learned journal (Marwick 1925). This was illustrated by a further photograph of the stone. The image is tightly cropped around the artefact, but the slab seems to have been placed on a wooden plank. A drystone wall is visible in the background. Although the photographer of this image is not known, the lighting, orientation, and composition, suggests that it is the work of local professional Tom Kent. A native Orcadian, Kent had emigrated to America in his teens and become apprenticed to the studio of society photographer M.J. Steffens in Chicago. He returned to Orkney in 1898 and set up a shop and studio of his own, specializing in a wide range of subjects including artefacts and archaeological sites (Tinch 1988: 182). Two further, attributed, prints by Kent of the Brodgar Stone survive in the Orkney Archives. It is interesting to compare these images with Leith's earlier photograph (Figures 7.1 and 7.2). Kent's studio for the shot, like Peter Leith's, was *en plein air*, in the farmyard at Brodgar, but the professional photographer substituted the roll of barbed wire for a wooden barrel. Kent lay the stone horizontally, but turned it 180 degrees from Leith's orientation. In one of the images (Figure 7.2), Kent

Figure 7.2 The Brodgar Stone, photographed by Tom Kent in 1926 (Orkney Library and Archives).

followed Leith in placing hammerstones from the discovery on top of the slab, but included a third stone found after the amateur had taken his photograph. But most significantly of all, in Kent's images the lines of the incised decoration are not chalked in. The early spring sunlight was at just the right angle to give the necessary contrast, raking across the edge of the slab and showing the carved lines perfectly (Thomas 2016: 3).

It was not until April 1925 that James G. Marwick, the local Provost, was able to visit Brodgar Farm to see Wishart's find. The place where the stone had been pulled from the field was no longer visible, but propped up against the wall in the barn, he found what he had come for: the large slab with the 'curious marks' that everyone had been talking about. Marwick immediately set about preparing his report on the discovery for the *Society of Antiquaries of Scotland.* Published in the Society's *Proceedings* (Marwick 1926), this article was to have a far wider readership, and influence, than the note on the discovery that had appeared the year before in the *Proceedings of the Orkney Antiquarian Society*. And crucially, it was Leith's image – rather than Kent's – which Marwick used to illustrate his report.

The year after the article's publication, and at James Marwick's suggestion, the Brodgar Stone was purchased for the National Museum in Edinburgh. It has been on display almost continuously since its accession. Although it was later placed vertically, the early museum display in the 1940s, and contemporary National Museum of Scotland photographs of the stone placed the stone horizontally on a plinth (Figure 7.3). This followed the mode of display seen in the first photographs of the stone by Leith and Kent. But despite having been on display in Edinburgh since the 1920s, in the wider archaeological literature, the

Figure 7.3 The Brodgar Stone on display in the Museum of Scotland, 1940s. The text reads: 'IA 44. Sculptured slab of sandstone having groups of chevrons, lozenges and oblique lines *incised* across one edge found beside two short cists at Brodgar, Stenness, Orkney. Purchased, 1927. Proc Vol LX, Page 35' (my emphasis; the reference is to Marwick 1926) (National Museums Scotland).

Brodgar Stone has received only passing attention. Consider how the Stone has been represented through drawn illustrations in the archaeological literature over the past four decades (Figure 7.4). Elizabeth Shee Twohig omitted it from her otherwise comprehensive 1981 work on megalithic and related art – although she briefly referred to it, unillustrated, in a 1997 article (Shee Twohig 1981, 1997). Colin Richards's doctoral thesis *An Archaeological Study of Neolithic Orkney* (1993) illustrated the decorated edge of the slab as a vertically oriented line drawing, following the orientation of the later museum display. In a subsequent analysis of 'Incised and Pecked Motifs in Orkney Chambered Tombs' (Bradley *et al.* 2000), the line drawing of the stone is oriented horizontally, in the same way as Peter Leith's original photograph. Alexandra Shepherd (2000) followed this format, but the motifs were abstracted from the outline of the stone (Thomas 2016: 191). Paul Brown and Graeme Chappell's (2005) illustration returned to the orientation of Tom Kent's photograph.

The slight variations between these drawn representations of the Brodgar Stone are superficially unimportant, but they reveal a number of more fundamental issues. Each drawing has been transcribed from a photograph, whether Leith's, Kent's, or the National Museum of Scotland's image (Figures 7.1–7.3). But the drawings also share something else in common: in each case the photographs have set the standard for not only what is depicted, but also what is actually *seen*. In particular, certain elements of the stone's style and motifs have been

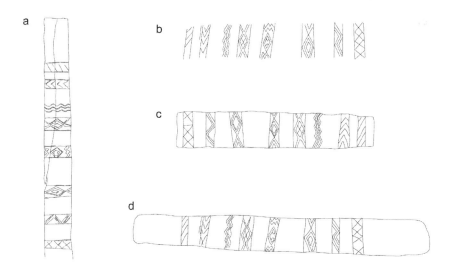

Figure 7.4 Line drawings of the Brodgar Stone. Drawn by Antonia Thomas, after (a) Richards 1993, 195, figure 8.13; (b) Shepherd 2000, 150, figure12.14a; (c) Brown and Chappell 2005, 40, figure 44.3; and, (d) Bradley *et al.* 2000, 61, figure 13.

entirely overlooked. Ever since Peter Leith chalked in the edge of the slab for his 1925 image, it has only been the stone's *incised* marks that have received attention. But there are also accompanying – and apparently overlying – *pecked* and *ground* marks along the stone's edge. A deep cup has been ground into the cross-and-lozenge band, and several distinct depressions can be seen by the banded lozenge design. These comprise three discrete peck-marks, arranged in a triangular manner, with a smaller peck-mark in the middle of one side (Figure 7.5a). Any possibility that these marks could be accidental, or due to damage, can be dismissed by comparison with other stones found at the Ness of Brodgar. The same 'triple-cup' motif, comprising three pecked, ground or drilled cups arranged in a triangular manner, occurs frequently throughout the assemblage. In the case of Small Finds SF17506, SF11546 and SF11566 (Figure 7.5b, c and d), there is a smaller peck-mark between two of the larger ones, as seen on the Brodgar Stone. In the case of SF11566 (Figure 7.5d), the pecked motif also overlies an incised banded design. Despite appearing to be part of a wider pattern of focused, deliberate marking, the pecked and ground working on the Brodgar Stone had, until recently, been entirely ignored (Thomas 2016: 193).

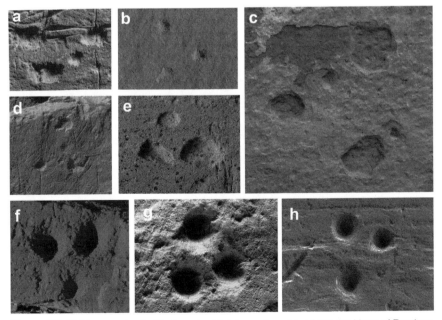

Figure 7.5 Eight examples (a–h) of pecked 'triple cup' marking from the Ness of Brodgar, comprising the Brodgar Stone and Small Finds (SFs) from excavations. From top left: (a) the Brodgar Stone; (b) SF17506; (c) SF11546; (d) SF11566; (e) SF16599; (f) SF6136; (g) SF7726; (h) SF11560 (photos: Antonia Thomas).

This simple recognition has profound implications. Through a range of incised, pecked and ground marks, the Brodgar Stone displays a sequence of decoration and alteration involving multiple engagements at different times. It is unclear that its decoration was ever 'finished', or that the Stone was ever intended to be a final form. Since its design would be added to, defaced, altered or augmented, any 'meaning' that we might seek to discern in the decoration cannot be fixed or static. These different acts of marking do not represent discrete 'events', but ongoing processes of differing durations, perhaps interrupted by long hiatuses or periods of intense activity. This challenges the Stone's status as art, and the purely visual consumption that this term implies. It leads us to consider the Brodgar Stone instead as a multi-durational artefact. It indicates that the *process* of working may be as significant as its *form* (Thomas 2016: 225). This has repercussions for how we understand the creation and appreciation of carvings in the Neolithic, and their relationship to different phases of architecture and activity.

This also tells us a great deal about the representation of the past in archaeology. The line drawings shown in Figure 7.4 depict only a virtual Brodgar Stone, as it appears to exist in various early photographs. But, and as Frederick Bohrer has argued, 'photography has . . . a double impact that does as much to create anew as to record what is pre-existent' (Bohrer 2011b: 32). The early photographs of the Brodgar Stone have effectively displaced the original artefact to become the *prima facie* evidence; what started as a representation has become the reality. This process is then continued as the line drawings themselves become represented as an objective point of fact, which is then interpreted in accompanying texts. As Stephanie Moser and Sam Smiles have observed,

> images of the past survive longer than the theories they were designed originally to support; *they linger on* in museum displays, as illustrations in archaeologically oriented books, and as part of popular culture.
>
> MOSER and SMILES 2005: 6; my emphasis

But the illustrations of the Brodgar Stone show only part of the picture. In each of the illustrations, the artefact has been presented as the result of a single event. Only one stage of decoration (the incising), and thus one imagined single point in time, has been represented (Thomas 2016: 192). The material past has been misread as if it is a fixed, *photographic* instant. And in this instance, the work of the camera, refracted through the drawn illustrations, is implicated in the representation of this moment in time. This demonstrates a wider problem of archaeological interpretation and recording that favours 'the moment of creation over the *duration of appreciation*' (Barrett 1999: 22; my emphasis).

I have argued that this problem relates to archaeology's loyalty to the linear time of *chronology*, which sees the past as a series of photographic *snapshots*. We might struggle to function without it (Lucas 2005: 27), but as an *unquestioned* and exclusive framework for understanding the past, chronology is clearly problematic (Thomas 2016: 184). We need to consider other ways of thinking about time.

Philosophy and time

How can philosophical approaches to time and duration contribute to these observations? As the 'single most pervasive component of our experience' (Sklar 1998: 413), time has often been conceptualized by philosophers as a tension between *stasis* and *change.* That idea is often illustrated by Zeno's famous paradox. This uses the example of an arrow in flight, which at any given moment, both moves and occupies a single point in time, thus demonstrating the fundamental *aporia* or paradox of time. It assumes that time is an infinite succession of instants, within which things cannot simultaneously occupy a point and change (Lucas 2005: 19–20). Aristotle refuted Zeno's paradox by arguing that time was not a series of instants, but rather an infinitely divisible continuum from past to future. This continuum was connected by the present as both a point and a line; for Aristotle, time was movement, *spatialized* to become an abstract and objective container for human action (Lucas 2012: 21).

At the start of the twentieth century, J.E. McTaggart's 'The Unreality of Time' (1908) revisited the *aporia* and distinguished between two types of time: A-series and B-series. In this model, A-series represents *time as continuum* (past-present-future), whilst the B-series emphasizes *time as a series of successive instants* (time as a sequence of historical dates). In the B-series, things always occupy the same relative position: they are earlier than/before, later than/after something else, and this quality is *permanent*. But in the A-series, time is problematic, as what was once future becomes present, and then past: it *changes* (McTaggart 1908: 458). Since something cannot be simultaneously past, present and future, the A-series has to be explained in terms of the B-series (Lucas 2005: 21). We do not experience time as the succession of points defining the B-series, so this experienced time defies simple measurement (Gosden 1994: 2). This tension between 'measured' and 'human' time lies at the heart of archaeological discussions of time, as we attempt to understand the human creation of time within a chronological framework created by measured time (Gosden 1994: 2; Thomas 2016: 184).

For McTaggart, the co-dependence of the A-series and B-series was contradictory, and so for him time must be *unreal* (McTaggart 1908: 458; Lucas 2005: 21). Following McTaggart, Henri Bergson also acknowledged the paradoxical nature of time, but without considering it any less 'real' (Bergson 1911). A major part of Bergson's thinking rests upon the suggestion that we do not experience time as a series of instants, but as *duration*, or *durée*. But what is significant to the current discussion, is the way in which Bergson's approach relates not only to the past, but also to the future; the way in which a consideration of duration

> involves the fracturing and opening up of the past and the present to what is virtual in them, to what in them differs from the actual, to what in them can bring forth the new.
>
> GROSZ 2005: 4–5

Many archaeologists have observed how Bergson's approach is particularly apposite when dealing with the multiple durations of past material remains. A Bergsonian approach to archaeology might recognize the material past not as a static sequence, but

> as intermingling remains that persist through time by virtue of qualities of durability. Every site, every place contains vestiges of its history, because the past, in its materiality, hangs on.
>
> SHANKS and SVABO 2013: 100

As Yannis Hamilakis and Jo Labanyi have noted, Bergson's ideas regarding duration are particularly pertinent when dealing with archaeological objects that 'were created at a certain point in time but have subsequently been reworked, re-engaged with and reactivated through human social practice', since such objects 'speak of time as coexistence rather than succession. And they embody, materially and physically, memory as duration' (Hamilakis and Labanyi 2008: 6). This understanding of the past as a *coexistence*, or *coalescence* of times (cf. Plummer 2012: 36) is pertinent to both archaeology and photography. It runs counter to thinking that sees the past as made up of static and discrete 'events', and instead suggests that the past is much less certain, and is rather 'like the future ...[existing] along a "spectrum of possibilities"' (Witmore 2012: 29).

Edmund Husserl subsequently developed Bergson's theories through his phenomenology of internal time-consciousness (Husserl 1966). Husserl argued that when we hear a musical tone, it flows; but when we represent it, we can only *represent* a series of notes. Similarly, he suggested, our consciousness perceives time as flowing, but we can represent this temporal flux only as a series of instants. Husserl thus presented a tension between the *representation* of time and the 'essential' character of time, with the former relegated to a secondary role (Lucas 2005: 22). Using Husserl's phenomenological theory of internal time consciousness as a starting point, Alfred Gell argued that our access to time is confined to the A-series flux, through which we interact with 'real' time, via the mediation of temporal maps which provide us with a surrogate for real time. These *reconstructions* of B-series time are not the real thing . . . but we are obliged to rely on them.

GELL 1992: 239–40; my emphasis

These observations have implications for how we understand the relationship between archaeology and the representation of the past.

Archaeology and representation

In archaeology, Husserl's tension emerges especially in relation to how time is *visualized* or *represented*; that is, how we 'reconstruct' B-series time. Artefacts, buildings, sites, and entire past activities and processes become preserved by record, supplanted by the 'secondary, mimetic topographies of fact and imagination: notebook, drawing, photograph, museum, archive' (Hicks 2016: 33). Three-dimensional artefacts are reconstructed and translated into photographs and two-dimensional decontextualized line drawings, which are easily printed and reproduced in book form (Bradley 1997). These allow the comparison of the form of objects, artefacts and buildings across a very wide geographical and temporal range, facilitating arbitrary judgements relating to aesthetic values or perceived evolutionary developments (Scott 2006: 637; Thomas 2016: 183). These practices of visual comparison relate to the overwhelming 'ocularcentrism' of Western science – the privileging of the visual over the other senses (Thomas 2008). More specifically, they form part of a wider anthropological tradition of representation, whereby the ability to visualize a society becomes synonymous with understanding it (Fabian 1983: 106).

The problem lies with the way in which archaeology's visual conventions still produce, and perpetuate, a particular conception of temporality. This is the time of chronology: linear, unidirectional and evolutionary. When represented as parts of a chronological sequence, the constituent features of an archaeological site are rendered atemporal (Chadwick 1998; Lucas 2005: 40). Archaeological discussions of carved stones such as the Brodgar Stone are frequently categorized and understood through groupings based on their visual characteristics. The stones are abstracted from their archaeological contexts, and reduced to formal qualities in order to be viewed and compared simultaneously (Thomas 2008: 8). Form is privileged over process. The past is objectified, seen, ordered, and *othered* through the lens of the archaeologist in the present (cf. Fabian 1983: 106; Thomas 2016: 183). Reduced to forms through photography and illustration, artefacts are removed from the processes by which they were made, placed and appreciated (Thomas 2016). They are presented as visual instants: defined by apparent similarities that reinforce their appearance as fixed, static entities, and reified to the point where they are uncritically accepted as a 'real' version of the past (Bailey and McFadyen 2010: 576). As Andrew Cochrane has pointed out, 'to end with a representational interpretation is understandable, to begin with one is problematic' (Cochrane 2018: 175). This is demonstrated by the early photographs of the Brodgar Stone: at first as representations of the past, these became reality; folded together with the object of the study itself and actively influencing what is *seen* (cf. Bradley 1997: 68). In this way, archaeology not only depicts time: it also creates it (González-Ruibal 2013: 14).

In reducing objects to instantaneous forms or *snapshots*, archaeology's reliance with time as chronology is inescapably bound up with how it makes time through representation. But thinking through photography can reveal something it shares with archaeology: the potential for approaching temporality and representation outside of this particular chronological framework – in particular to think about duration.

Photography and duration

Photography has always had a 'deeply ambivalent relationship to time' (Hunt and Schwarz 2010: 259). From the outset, with Daguerre and Muybridge's early images of objects and bodies in flight, photographs have seemed to isolate things and people from the flow of time, in a 'kind of visual aporia' (Lowry 2010: 54). With the birth of cinema, the moving image brought 'a fundamental change in our understanding of time, so much so that photographic and cinematographic metaphors came to illustrate Bergson's own philosophy of perception' (Sutton 2009: 69). The different durational qualities of the moving image have been a common theme in discussions of the temporality of photography (Plummer 2012: 1). Gilles Deleuze, for example, used Bergson's account of duration in his account of the 'movement-image' and 'time-image' in cinema (Deleuze 1989a, 1989b). Deleuze's distinction between the idea of film as a series of successive images, or 'a coexistence of distinct durations' (Deleuze 1989b: xii) – echoes the distinction explored above between time as a series of successive instants and time as a simultaneous continuum. But just as in the study of cinema, so in the study of still photographs we can explore the durational, as well as purely the instantaneous, dimensions of the image. A photographic exposure may be made 'in the blink of an eye', but the photograph is made to outlast that instant, with future experiences of looking at the photographic image belonging to 'the subjective register of *durée*' (Burgin 2010: 131). Through the Bergsonian account of time and Deleuze's analysis of cinema, we might understand photographs as multi-durational artefacts (Plummer 2012: 36). As Victor Burgin has observed we expect to be told the running time of a film or video; we do not normally ask: 'How long is the photograph?' – but the question is not entirely irrelevant'.

BURGIN 2010: 131

This chapter started with a photograph (Figure 7.1). But what durations are manifested in Peter Leith's image of the Brodgar Stone? There is, of course, the distant Neolithic when the stone was originally carved and used. Running through the image is also the 1920s when the stone re-emerged to be found and photographed; the low winter light of the afternoon February sun that illuminates the photograph; the moment that Leith released the camera's shutter and the exposure time of the plate itself. Each make their own contribution to its biography, but as we have seen, there are also further durations at play. Multiple processes and timescales contribute to the biography of the stone up until the point when it was found, but there is also the continuing story and 'afterlife' of both the image and the artefact. These endure long after the shutter was released. Time operates here not as linear chronology, but as the effect of coexisting and overlapping durations (Deleuze 1989b: xii).

As a time-image, Leith's photograph transcends chronological time and produces instead a coexistence of past, present and future, as the past of the recorded event is fused with the present tense of its viewing and the afterlife of the image itself. Photography disrupts 'our common-sense understanding of the relationship between past and present' (Lowry 2006: 65). It not only depicts the past, but also points to the future: 'it not only tells us *what has been* but what *will* be' (Plummer 2012: 2, original emphasis). This encapsulates the multiple durations of photography, and the ability of photographs to 'condense different times as coexistence rather than succession; for example, the having-been-there of the time when the photo was taken, and the here-and-now of us viewing it, and often several other times in between' (Carabott *et al.* 2015: 10). As with the Deleuzian notion of the time-image, so with the biography of the Brodgar Stone: 'the sheets of past coexist in a non-chronological order' (Deleuze 1989b: xii).

This understanding serves to challenge photography's position of assumed objectivity (cf. Hunt and Schwarz 2010). If the past is not fixed, then it is not possible to capture a fixed moment from the past. It reminds us that

the only place the past can exist is in the present. . .[It] does not exist in actuality, but is the *virtual* form of the past, accessible through various practices of remembering.

HODGES 2008: 411

This allows us to see photography as 'a complex and embodied cultural process of which the photographs themselves are only the final outcome' (Edwards 2014:

179). By recognizing their multi-durational character in this way, photographs become more than just visual. They cease to be fixed, neutral or objective records, or the after-effects of cultural history. They become fluid and mutable, transcending temporal boundaries as if they were memories (Shevchenko 2014: 4). They are *social* objects made meaningful through different forms of apprehension (Edwards 2005: 27). And as we have seen in Peter Leith's image of the Brodgar Stone, they are also temporal interventions that create multiple durations spanning pasts, presents and futures.

Thursday 14th November 2013. Langbigging, Stenness, Orkney.
Peter Leith's son, also called Peter, had raised an Orkney flag on the pole in his garden to help me find his house. The building was unexpectedly modern, but the land had been occupied by the Leith family for generations. He opened the door and showed me to the sitting room. On a round table in front of the sofa, he had laid out an A4 folder containing various photographs, hand-written notes and newspaper cuttings. Later that afternoon, two cigar boxes appeared, each containing assemblages of steatite spindle whorls, stone tools, flint arrowheads and polished stone axes. These items from the Norse period, the Iron Age and the Neolithic told stories from the 1920s, the 1940s, and after – stories of collecting and curation. Now in his late eighties, some were collected by Peter as a boy; others had been given to him by his father.

I had written to Peter about the Brodgar Stone the week before. It was his father's photograph which had illustrated James Marwick's paper 'Discovery of Stone Cists at Stenness, Orkney' in the Proceedings of the Society of Antiquaries of Scotland *in 1926, and I was keen to understand more about it. Peter told me about James Wishart, and about his father: how he had chalked in the lines on the stone to make them stand out. Holding the photograph and talking to Peter, I learned more about the image and how it had come to be taken, but I also learned more about the Neolithic artefact. I have long been fascinated with both this stone and its reproduction, the image and the artefact, and I felt honoured when Peter gave the print to me. For Peter, this was also a very personal possession: a family photograph, which told his father's story of an image taken two years before he was born. More than just an image, this had become an artefact itself; a multi-durational object which linked the past, present, and future.*

Conclusion

This chapter has explored the idea that both archaeology and photography engage with and create multiple durations. I have argued that a consideration of photography as multi-durational has repercussions for how we think about time, and the ontological status of the past, in archaeology. Through the photograph of the Brodgar Stone taken by Peter Leith in 1925, this chapter has considered the lasting traces of Neolithic incised and pecked decoration redrawn in 1920s chalk, and the enduring effects of a shutter released in Orkney, on a bright February afternoon over 90 years ago.

We know today that the Brodgar Stone had a complex biography in the Neolithic, comprising several stages of marking and alteration. However, only one of these stages – the bands of incised decoration – has received attention since its discovery. Leith's chalked-in photograph played a central role in this narrative. Through these chalked lines, the conventions of chronological, successive time favoured, and perpetuated, by normal archaeological recording practices presented the object as the product of a single Neolithic carving event. Other durations were erased, and new durations introduced. As an archaeological interpretation, the photograph represented a further temporal intervention, enacting the singular and static interpretations offered for Neolithic art and architecture in Orkney. 'Frozen in time' at a particular point in its life, the nuances of the stone's biography have been erased and ignored. Its early photographic representations became artefacts themselves – the *prima facie* evidence from which archaeologists created their line drawings and narratives (Thomas 2016:

Figure 7.6 Peter Leith with his father's photograph in 2013 (Photo: Antonia Thomas). Reproduced with the kind permission of Peter Leith.

227). But in the case of Leith's photograph of the Brodgar Stone, we have seen that the image is not simply a medium for objective representation, and nor is it just a 'final outcome'.

My meeting with Peter Leith junior brought another intervention in endurance and memory (Figure 7.6). Holding the print of his father's photograph in his hand, the effects of duration were tangible in the faded and torn photographic paper, stained and yellowed on the back where someone had once written *Stone found at Broadgar* [sic] (Figure 7.7). As a physical object, the photograph had been only occasionally viewed in the 90 years since it was developed. It will, like the Stone itself, eventually decay. But the material past endures through its representation; it is made durable through 'iterative association and continued regeneration' (Witmore 2012: 29). Running in parallel with the photograph's entropic duration are other temporalities. These include the subjective duration of the viewer (cf. Burgin 2010: 131); its living on through secondary re-drawings and re-publications (including those in this chapter); and the reflection of these durations back from the photograph to alter the biography of the stone itself. Time here is non-linear. The photograph encompasses not just 1920s fieldwork and Neolithic stone-carving, but also contemporary knowledge.

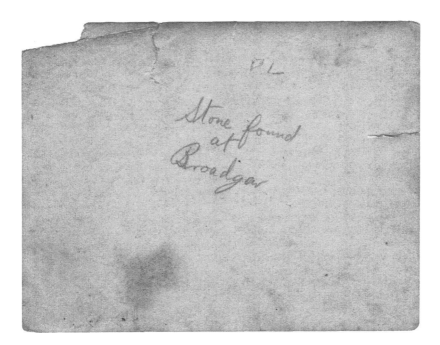

Figure 7.7 Reverse of Peter Leith's photograph of the Brodgar Stone (Photo: Antonia Thomas). Reproduced with the kind permission of Peter Leith.

As image, as object, and as idea, the Brodgar Stone cannot represent a snapshot of a fixed moment in time. It is a multi-durational artefact. Both Leith's photographic print and the Brodgar Stone as a museum object traffic between the multiple durations of the Neolithic, the 1920s, and my 2013 meeting with Peter Leith junior, whilst the image's *future* has affected the *past* of the artefact. In this way, photography provides a paradigm for archaeological intervention (Shanks and Svabo 2013). As durational image and durational artefact, and as knowledge in visual and material form, through time the photograph and the Stone decompose one another.

References

Badger, G. 2007. *The Genius of Photography: How Photography Has Changed Our Lives*. London: Quadrille Publishing.

Baetens, J., A. Streitberger and H. van Gelder 2010. Introduction. In J. Baetens, A. Streitberger and H. van Gelder (eds) *Time and Photography*. Leuven: Leuven University Press, pp. vii–xii.

Bailey, D. and L. McFadyen 2010. Built Objects. In D. Hicks and M.C. Beaudry (eds), *The Oxford Handbook of Material Culture Studies.* Oxford: Oxford University Press, pp. 562–587

Barrett, J.C. 1999. The Mythical Landscapes of the British Iron Age. In W. Ashmore and A.B. Knapp (eds) *Archaeologies of Landscape.* Oxford: Blackwell, pp. 253–265,

Bergson, H. 1911. *Creative Evolution* (trans. A. Mitchell). New York: Henry Holt.

Bohrer, F. 2005. Photography and Archaeology: The Image as Object. In S. Smiles and S. Moser (eds) *Envisioning the Past: Archaeology and the Image*. Oxford: Blackwell, pp. 180–191.

Bohrer, F. 2011a. *Photography and Archaeology*. London: Reaktion Books.

Bohrer, F. 2011b. Archaeology, Photography, Sculpture: Correspondences and Mediations in the Nineteenth Century and Beyond. In P. Bonaventura and A. Jones (eds) *Sculpture and Archaeology*. Farnham: Ashgate, pp. 31–44.

Bradley, R. 1997. 'To See Is to Have Seen': Craft Traditions in British Field Archaeology. In B.L. Molyneaux (ed.) *The Cultural life of Images: Visual Representation in Archaeology.* London: Routledge, pp. 62–72.

Bradley, R., T. Phillips, C. Richards and M. Webb 2000. Decorating the Houses of the Dead: Incised and Pecked Motifs in Orkney Chambered Tombs. *Cambridge Archaeological Journal* 11(1): 45–67.

Brown, P.M. and G. Chappell 2005. *Prehistoric Rock Art in the North York Moors.* Stroud: Tempus.

Burgin, V. 2010. The Eclipse of Time. In J. Baetens, A. Streitberger and H. Van Gelder (eds) *Time and Photography*. Leuven: Leuven University Press, pp. 125–140.

Carabott, P., Y. Hamilakis and E. Papargyriou 2015. Capturing the Eternal Light, Photography and Greece, Photography of Greece. In P. Carabott, Y. Hamilakis and E. Papargyriou (eds) *Camera Graeca: Photographs, Narratives, Materialities*. Farnham: Ashgate, pp. 3–21.

Chadwick, A.M. 1998. *Archaeology at the Edge of Chaos: Further Towards Reflexive Excavation Methodologies*. Assemblage 3, 18th January 1998. Available at https://

archaeologydataservice.ac.uk/archives/view/assemblage/html/3/3chad.html (accessed 28th March 2019).

Cochrane, A. 2018. Archaeology through the Looking Glass: Photographic Documentation and the Politics of Display. In A.M. Jones and A. Cochrane *The Archaeology of Art: Materials, Practices, Affects*. London and New York: Routledge, pp.173–182.

Deleuze, G. 1989a. *Cinema 1: The Movement-Image* (trans. H. Tomlinson and R. Galeta). Minneapolis: University of Minnesota Press.

Deleuze, G. 1989b. *Cinema 2: The Time-Image* (trans. H. Tomlinson and R. Galeta). Minneapolis: University of Minnesota Press.

Edwards, E. 2005. Photographs and the Sounds of History. *Visual Anthropology Review* 21 (1/2): 27–46.

Edwards, E. 2011. Tracing Photography. In J. Ruby and M. Banks (eds) *Made to Be Seen: Perspectives on the History of Visual Anthropology*. Chicago: University of Chicago Press, pp. 159–189.

Edwards, E. 2014. Out and About: Photography, Topography, and Historical Imagination. In O. Shevchenko (ed.) *Double Exposure: Memory and Photography*. New Brunswick: Transaction Publishers, pp. 177–209.

Fabian, J. 1983. *Time and the Other: How Anthropology Makes Its Object*. New York: Columbia University Press.

Gell, A. 1992. *The Anthropology of Time: Cultural Constructions of Temporal Maps and Images*. Oxford: Berg.

González-Ruibal, A. 2013. Reclaiming Archaeology. In A. González-Ruibal (ed.) *Reclaiming Archaeology: Beyond the Tropes of Modernity.* Oxford: Routledge, pp. 1–29.

Gosden, C. 1994. *Social Being and Time.* Oxford: Blackwell.

Grosz, E. 2005. Bergson, Deleuze and the Becoming of Unbecoming. *parallax* 11(2): 4–13.

Hamilakis, Y. and F. Ifantidis, 2015. The Photographic and the Archaeological: The 'Other' Acropolis. In P. Carabott, Y. Hamilakis and E. Papargyriou (eds) *Camera Graeca: Photographs, Narratives, Materialities*. Farnham: Ashgate, pp.133–157.

Hamilakis, Y. and J. Labanyi 2008. Introduction: Time, Materiality and the Work of Memory. *History and Memory* 20(2): 5–17.

Hamilakis, Y., A. Anagnostopoulos and F. Ifantidis 2009. Postcards from the Edge of Time: Archaeology, Photography, Archaeological Ethnography (A Photo-Essay). *Public Archaeology: Archaeological Ethnographies* 8 (2–3): 283–309.

Hicks, D. 2016. Meshwork Fatigue. *Norwegian Archaeological Review* 49(1) 33–39.

Hodges, M. 2008. Rethinking Time's Arrow. Bergson, Deleuze and the Anthropology of Time. *Anthropological Theory* 8(4): 399–429.

Hunt, L. and V. Schwarz 2010. Capturing the Moment: Images and Eyewitnessing in History. *Journal of Visual Culture* 9(3): 259–271.

Husserl, E. 1966. *The Phenomenology of Internal Time-Consciousness* Bloomington: Indiana University Press.

Lowry, J. 2006. Portraits, Still Video Portraits and the Account of the Soul. In D. Green and J. Lowry (eds) *Stillness and Time: Photography and the Moving Image*. Brighton: Photoworks/Photoforum, pp. 65–78.

Lowry, J. 2010. Modern Time: Photography and the Contemporary Tableau. In J. Baetens, A. Streitberger and H. van Gelder (eds) *Time and Photography*. Leuven: Leuven University Press, pp. 47–64.

Lucas, G. 2005. *The Archaeology of Time*. London: Routledge.

Lucas, G. 2012. *Understanding the Archaeological Record.* Cambridge: Cambridge University Press.

Marwick, H. 1925. Note on Incised Stone Found at Brodgar. *Proceedings of the Orkney Antiquarian Society* 3(Session 1924–1925): 91.

Marwick, J.G. 1926. Discovery of Stone Cists at Stenness, Orkney. *Proceedings of the Society of Antiquaries of Scotland* 60: 34–36.

McFadyen, L. 2013. Designing with Living: A Contextual Archaeology of Dependent Architecture. In B. Alberti, A.M. Jones and J. Pollard (eds) *Archaeology after Interpretation: Returning Materials to Archaeological Theory.* Altamira: Left Coast Press, pp. 135–150.

McTaggart, J.E. 1908. The Unreality of Time. *Mind* 17(68): 457–474.

Moser, S. and S. Smiles 2005. Introduction: The Image in Question. In S. Smiles and S. Moser (eds) *Envisioning the Past: Archaeology and the Image*. Oxford: Blackwell, pp. 1–12.

Olivier, L. 2001. Duration, Memory, and the Nature of the Archaeological Record. In H. Karlsson (ed.) *It's About Time: The Concept of Time in Archaeology.* Goteborg: Bricoleur Press, pp. 61–70.

Plummer, S. 2012. Photography and Duration: Time Exposure and Time-Image. *Rhizomes* 23 www.rhizomes.net/issue23/plummer/index.html (Accessed 15 May 2017).

Richards, C. 1993. *An Archaeological Study of Neolithic Orkney: Architecture, Order and Social Classification*. Unpublished Ph.D. Thesis, University of Glasgow.

Scott, S. 2006. Art and the Archaeologist. *World Archaeology* 38(4): 628–643.

Shanks, M. 1997. 'Photography and Archaeology. In B.L. Molyneaux (ed.) *The Cultural life of Images: Visual Representation in Archaeology.* London: Routledge, pp. 73–107.

Shanks, M. and C. Svabo 2013. Archaeology and Photography: A Pragmatology. In A. González-Ruibal (ed.) *Reclaiming Archaeology: Beyond the Tropes of Modernity*. London: Routledge, pp. 89–102.

Shee Twohig, E. 1981. *Megalithic Art of Western Europe.* Oxford: Clarendon Press.

Shee Twohig, E. 1997. 'Megalithic Art' in a Settlement Context: Skara Brae and Related Sites in the Orkney Islands. *Brigantium* 10: 377–389.

Shepherd, A. 2000. Skara Brae: Expressing Identity in a Neolithic Community. In A. Ritchie (ed.), *Neolithic Orkney in its European Context*. Cambridge: McDonald Institute Monographs, pp. 139–158.

Shevchenko, O. 2014. Memory and Photography: An Introduction. In O. Shevchenko (ed.) *Double Exposure: Memory and Photography*. New Brunswick: Transaction Publishers, pp. 1–17.

Sklar, L. 1998. Time. In E. Craig (ed.) *The Routledge Encyclopedia of Philosophy.* London: Routledge, pp. 413–417.

Sutton, D. 2009. *Photography, Cinema, Memory: The Crystal Image of Time*, London: University of Minnesota Press.

Szarkowski, J. 1980. *The Photographer's Eye*. London: Secker and Warburg.

Thomas, A. 2016. *Art and Architecture in Neolithic Orkney: Process, Temporality and Context*, Oxford: Archaeopress.

Thomas, J. 2008. On the Ocularcentrism of Archaeology. In J. Thomas and V.O. Jorge (eds) *Archaeology and the Politics of Vision in a Post-Modern Context.* Newcastle: Cambridge Scholars Publishing, pp. 1–12.

Tinch, D.M.N. 1988. *Shoal and Sheaf: Orkney's Pictorial Heritage*. Belfast: Blackstaff Press.

Witmore, C. 2012. The Realities of the Past: Archaeology, Object-Orientations, Pragmatology. In B. Fortenberry and L. MacAtackney (eds) *Modern Materials: The Proceedings of CHAT Oxford, 2009*. British Archaeological Reports International Series 2363. Oxford: BAR Oxford Ltd.

Wollen, P. 1984. Fire and Ice. *Photographies* 4: 118–120.

8

PHOTOGRAPHING BUILDINGS

James Dixon

This chapter discusses the photographic record of two terraces of early nineteenth-century housing either side of a public house in the London Borough of Southwark. My involvement was as a field archaeologist recording the buildings in 2015 in advance of their redevelopment, work undertaken by Museum of London Archaeology (MOLA) in accordance with planning regulations – the kind of photographic work undertaken by professional archaeologists all the time. The terraces had accommodated ground floor businesses until a few years before, and the pub was in operation until around 2008. Many of the upper floors of the terraced buildings had been disused since the 1970s. The buildings have now been entirely refurbished for use by a local educational institution.

The archive was more revealed than created. My colleague and I had tried to make it as small as possible. Despite taking hundreds of photographs of the buildings, we were joined on site for one afternoon by a professional architectural photographer whose job it was to replace our hurried snaps with better quality photographs for the archive (Figures 8.1 and 8.2). So far, so neat and tidy. Two sets of photos – one official, for the future, the other not, for the present. But as the project drew to a close, the manager of the refurbishment works taking place on site at the same time as our recording work asked me, 'Would you like a copy of all my weekly report photos?' Well, yes. So the archive expanded, including now a vast collection of close-up shots of repair works and test excavations (the non-archaeological kind) (Figure 8.3). And that wasn't all. Dotted around the building had been floor plans with people's names written on them in this corner or that. Back in the office, having retrieved these maps from disposal, I looked up the names and emailed those whom I could locate online to ask what they had

* Thank you to Azizul Karim and Maggie Cox for work on site and to Louise Davies who managed the work, and to David Bowsher (MOLA) and Cathy Gale (artist) for allowing use of photographs.

to do with the building. A few replied. They were art students and the building had hosted a guerrilla exhibition during its years of disuse. 'Could I see any photos?', I asked. Most couldn't help; photos lost, files won't open, no longer an artist. . . But Cathy Gale had photos, and so the archive expanded again (Figure 8.4).

Where is the archive of the transition of these buildings? A handful of photographs by a professional photographer is archived along with the archaeological report on the site, deposited with the London Archaeological Archive and Research Centre (LAARC) in Hackney. But there are also many other photographs, made through archaeological recording but also through interventions in the buildings' transformation by contractors and artists, spread across computer hard drives and mobile phones, for now at least. This chapter aims to take the second body of digital photographs seriously as a kind of archive made, like the official record of professional images, during these buildings' period of change, documenting their existence between formal inhabitation and use, as an unpopulated space. This secondary, informal archive contrasts with the official archive's representation of the earliest elements of the building, those considered of historical interest and architectural value, which adopts a kind of timeless approach to architectural beginnings rather than afterlives. Does the attention of this informal archive to buildings as ongoing projects, rather than fragments of original architectural intentions and achievements, address more adequately how we think about buildings as archaeologists today? With its focus on the building as a trace to be fixed with the camera and preserved in the museum, I want to suggest that the professional archive is not just out of step with how we think about Archaeology and Architecture – but also with how we think about Archaeology and Photography.

*

A formal archive is representational, inasmuch as it makes selections of what to include and what to omit. Attending to the informal archive – the official plus the discarded plus the found – has the potential to disrupt this representational process, and to bring the archive closer in form to what it documents – a form which, in this instance and in most other examples of commercial archaeology, is one of ongoing transformation. There are parallels here with anthropological thinking about architecture, which is increasingly adopting longer temporal frames in understanding built environments (Buchli 2013: 47), and in some pioneering developments in professional archaeology to document contemporary life as well as the distant past, such as Emma Dwyer's study of the London Overground East London line (Dwyer 2009, 2015).

Fadwa El Guinidi describes visual anthropology as a practice that is 'anthropology conceptually, and ethnography empirically' (El Guinidi 2015; 442). Following Buchli and Dwyer, we might imagine that Visual Archaeology in situations of the recording of a building undergoing transformation can bring both Archaeology and Ethnography into play in a similar manner – *empirically archaeological* in that it empirically documents the material evidence of major phases of construction, use, abandonment, demolition, etc.; and *conceptually ethnographic* in that it observes and participates in the ongoing human uses of contemporary spaces, in which temporal depths take different forms: inhabitation, routine, memory, bureaucracy, etc. Such a Visual Archaeology engages not just with architectural ruins and remains, but experiments with what it means to photograph buildings in an explicitly archaeological way by attending to the in-between times and spaces of transition and fleeting human presences – finding the dispersed visual archive of others as well as making a record ourselves.

To what extent might we draw our theoretical approaches to archaeological photography into a closer relationship with how we study the built environment? A range of studies have shown how archaeological approaches to buildings go beyond architectural history and a focus on the material fragment to study the lived durations of buildings (Bailey 1990; Bailey and McFadyen 2010; Edensor 2011; Green and Dixon 2016; Hicks and Horning 2006). But at stake here is also a broader set of questions about the relationships between method and theory, and the way in which the institutional relationships between commercial and university-based archaeology, in the UK and beyond, has mapped onto ideas about this distinction. Over the past two generations, this mapping has played a central part in the assumption that the collection and archiving of data can be distinguished from wider interpretive and synthetic endeavours.

The mapping has a primary physical, geographical dimension in that archaeologists based in commercial companies and in universities almost always study, excavate and analyse different sites and assemblages from each other – under different circumstances, with different justifications and agendas, and in different areas or regions. Commercial buildings archaeology is almost always a form of mitigation, recording a building before it is demolished or otherwise hugely altered. The immense amounts of data and documentation, including hundreds of thousands of photographs taken every year and archived in museum collections, are almost never subject to further analysis. Developments in method and theory in each field rarely cross-fertilize, and the wider interpretation of commercial archaeology sites remains, practically, with commercial archaeologists themselves. Since this is the case, a major challenge is for commercial archaeology to find innovative ways for their fieldwork to result in more than just the collection of data or the recording of buildings – to find new ways of interpreting through fieldwork at the huge numbers of sites that comes with most commercial archaeological practices. Reimagining archaeological photography of standing buildings, as a method and as a genre, represents a central element of that challenge – and one that would simultaneously allow university-based studies to engage at a new scale with the masses of sites and buildings that it currently neglects.

One thing that unites the vast majority of buildings archaeology, whether in higher education or commercial environments, is that it operates in total isolation from anthropological questions (Johnson 1993; viii–ix), focusing narrowly on material remains alone. It is envisioned as providing additional primary data for architectural history, rather than offering any distinctive perspective on human lives in built environments. The focus is therefore often guided by periodization and the idea that a building is of particular value or interest in relation to a time, architect or style. Archaeology's unique ability to document how buildings come into and out of existence, becoming intertwined with human lives as they change in use and meaning, is often neglected (Bailey 1990; Cairns and Jacobs 2014). The collision of scales of architectural, human and community 'life', from conception to disaggregation, complicates our conception of historical time-lines and human biographies as linear chronologies. Disciplinary interventions imagine and occur at different times in a building's existence; the architectural historian's interest in the drawings of an architect contrasts with the commercial archaeologist's physical and conceptual point of entry, which is usually somewhere between de-occupation and demolition, mitigation in advance of loss. A major problem for commercial archaeology begins here, since it relies on art history for its main interpretive framework, and thus finds itself out of step with the time in which it encounters and intervenes with the building. This difference in conceptions of the time of a building – the fragmentary remains of its origins and its contemporary, ongoing state – is what led to the different photographic practices at Southwark between ourselves as fieldworkers and the official photographer.

A series of guidance documents have advised on the recording of buildings in UK archaeology since the 1990s, including the Royal Commission for the Historical Monuments of England document *Recording Historic Buildings: A Descriptive Specification* (English Heritage 2006), and Historic England's more recent guidance document *Understanding Historic Buildings: A Guide to Good Recording Practice* (Historic England 2016). Such guidance describes photography as one among other field recording techniques in buildings archaeology, like measured drawn survey.

> Like drawings, photographs amplify and illuminate a record. In many cases they are a more efficient way of capturing data than either drawings or written description, but they also valuably supplement and verify drawn or written records . . . Photography is generally the most efficient way of presenting the appearance of a building, and can also be used to record much of the detailed evidence on which an analysis of historic development is based. It is also a powerful analytical tool in its own right, highlighting the relationships between elements of a building and sometimes bringing to light evidence which is barely registered by the naked eye . . . In record photography the needs of the record should be paramount, but pictorial qualities, which often give life and meaning to architectural forms, should

not be neglected. Photographs which aim to convey the 'atmosphere' of a building, typically using available light, can be especially evocative, but should form a supplement to, not a substitute for, a series of well-lit images.

Historic England 2016: 17–20

Here the idea of the buildings archaeologist as gathering raw data in a fragmentary form for interpretation for the purposes of architectural history, or conservation or restoration, at a later date, away from the field, is clearly expressed. Materials, styles, techniques of construction are treated, like the question of how a building stands up (Salvadori 1991), as matters of fact. Just as they draw elevations, plans and sections, so they photograph elevations and internal spaces to mimic such visual analysis, with rectified photography and photogrammetry even producing photographs that are almost as accurate as scaled metric survey. Any technique of recording 'atmosphere' (Cole 2017) is decidedly secondary to documenting stonework, bricks and mortar. Any sense of commercial buildings archaeology, including buildings photography, as something more than a visual record and an archive – a mode of analysis and a situated intervention in the context of transformation – is missing.

One field in which an influential paradigm of the idea of the contemporary nature of photographing buildings has developed is in 'ruin photography' – a development with increasing influence upon Archaeology (see Pétursdóttir and Olsen 2014). The focus is on abandoned spaces, where the departure of people is accompanied by the physical deterioration of the built environment (Edensor 2011). The decay of domestic spaces like living rooms and bedrooms, public spaces like shopping malls or cinemas, and industrial spaces like empty factories are used to produce a distinctive aesthetic. Will Ellis's photo series on Sea View Children's Hospital on Staten Island (Ellis 2015) is typical. Technically, in terms of framing and light, they hold much in common with professional archaeological photographs of buildings. But considered together, the frequent inclusion of furniture like chairs and beds, and other personal items, for example in an image of piled-up patient records in the abandoned hospital, operate to evoke human absence and the time of inhabitation and absence in a way that the architectural space alone does not.

The lack of such material evocations of human occupation in most archaeological photography of buildings, due to the circumstances of most such fieldwork, contrasts with the archaeological aide-memoire photographs discussed here in the case study on the Southwark terraces. The artist's exhibition photos also play a particular role in rehumanizing the empty spaces of archaeological intervention. One contains images of real people, the other documents an individual's investigation of the space. Both put people back into what are, if best practice is adhered to, empty rooms within architectural structures. These concerns are taken further in the photographic practices of 'urban exploration'. The presence of the photographer, as an 'explorer' who transgresses boundaries (Garrett 2013) is much stronger than in archaeological photography of buildings – whether behind the camera or appearing in the frame.

*

With these themes in mind – the potential of buildings photography as a mode of analysis and intervention rather than simply archaeological documentation, and as a practice that can foreground human life as well as material detail, extending from those who have dwelled in a building in the past to the fieldworker in the present – let us return to the detail of the four photographic archives of the Southwark terraces.

Photographer 1: Architectural

The principal architectural photography on site was that undertaken by the Museum of London Archaeology photographer, and the images produced forms

Figure 8.1 Photographs of nineteenth-century buildings at Southwark, taken for archiving by an archaeological photographer, 2015 (Museum of London Archaeology).

the primary visual archive of the site. In the case of this particular set of buildings, access across the whole site was relatively restricted and the survival of original fabric and fixtures partial only. As the buildings were terraced and all of the same plan, it was decided to try to make up a single representative building through photography, as well as photographing other features and spaces where possible. This resulted in the recording, over the whole site, of a minimum of one full building. Part of the results of this exercise is shown in Figure 8.1. These photographs are a more time-consuming undertaking than any of those to follow. Typically, these official photographs require a number of personnel and in this case involved the MOLA photographer, an archaeologist advising on the content of individual shots, and one of the contractor's site staff assisting with lighting requirements and access. An effort was made in each shot to record as much of any room as possible, which in practice means including the majority of at least two elevations in each photograph as well as elements of the floor and ceiling. Ranging poles for scale are not usually used in this kind of archaeological building recording (although this is ultimately a preference of individual archaeologists and photographers) as the standardizing of building fabric renders them unnecessary in most cases (photographs of architectural detailing may differ). The set comprises 64 images.

Photographer 2: Aide-memoire

The second set of photographs are those that I took myself as a buildings archaeologist on site while going from room to room producing the written description of the structure, its spaces and surviving historic features (Figure 8.2). There is not necessarily an effort made to systematically record the entirety of any built space, as what is included in these photographs depends on the working methods of an individual archaeologist. In this case, the photographs generally reflect the typical circulation through the spaces of the building and follow the order of a comprehensible building description, with photographs taken of architectural details where they are likely to add to the architectural interest or understanding of the building. The photographs are not taken with a tripod and a flash was only used when absolutely necessary, resulting in variable quality across the whole set of photographs although the detail of images is generally clear and in focus. The images are not systematic. Across the whole set of 497 images, the movement through the site is clear, with images of spaces and architectural details punctuated with images giving the location of the work at any given time, in this case property numbers painted on the rear doors in red paint. Whereas the official archive photographs visually describe the representative building as a structure, as a piece of architecture, the aide-memoire photographs describe the access around the site and the imposition of the contemporary building site onto the historic buildings. What is notable about these photographs, aside from the volume, is that they were

Figure 8.2 Aide-memoire photographs of nineteenth-century buildings at Southwark taken by a buildings archaeologist, 2015 (Museum of London Archaeology).

taken with the specific aim of re-rendering their contents as text, in the form of a structural description. They exist specifically not to be used. Yet, they clearly show something different to the official images in Figure 8.1; the contents of the part-cleaned, part-cluttered rooms, the circulation between spaces, the process of archaeological investigation itself.

Photographer 3: Contractors

The final set of photographs taken through the period of restoration of the buildings is those taken by the contractor's project manager to record daily

Figure 8.3 Contractor's work-in-progress photographs of nineteenth-century buildings at Southwark, 2015.

progress of work (Figure 8.3). These are notable in not aiming to capture any structural detail of the building per se, but the processes involved in altering the overall structure. In this case, that mainly included the digging of test pits in the building basements and the removal of architectural elements for refurbishment and reinstatement. These photographs have been included as they demonstrate the process of change taking place in the building in a way that the archaeological photographs do not. Typically, archaeological photography will happen on a site in advance of the commencement of development and so no trace of the development itself will be seen. Occasionally site work will include watching brief work or, as in this case, work alongside contractors, and in these cases the process of change may be visible in the archaeological record. However, in general this represents a record of a process wholly absent from the archaeological archive. The photographs themselves are of a similar technical quality to the aide-memoire photographs of the archaeologists, although are more repetitive and record in depth the progress of very small interventions such as test pits or the cleaning of architectural elements. As they are attached to emails or filed on a daily basis, there is no attempt to archive the images for later interpretation.

Photographer 4: Artists

The final set of photos comes from an artist involved in a guerrilla exhibition held in the public house on the site while it was out of use in 2009 (Figure 8.4). Evidence of the exhibition was clear during the site survey, not least because of

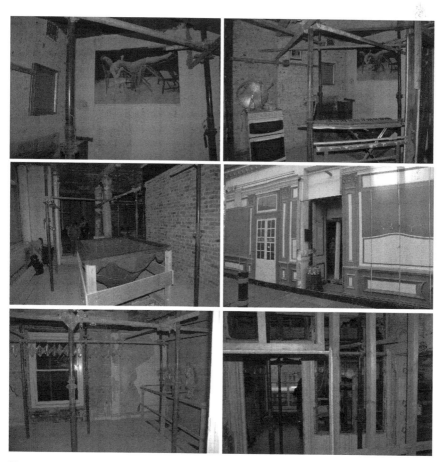

Figure 8.4 Artists' photographs of an exhibition in nineteenth-century buildings at Southwark, 2009 (photos: Cathy Gale).

maps of the floors of the building pinned to the walls on each floor with names written over them explaining whose work was installed where. As mentioned above, I retained a number of these maps and used them to contact the artists involved in the exhibition, a number of whom were happy to share photos of and information about the exhibition. The photos capture a period of time not usually documented archaeologically – moments of use during a time in which the building was officially unused. They exist somewhere between the photographs of the building in use that we find in the local archive and the archaeological recording that aims to take place between de-occupation and the commencement of refurbishment or demolition. The photos are important, not only for the reasons just outlined, but because they contain people and artworks. They demonstrate multiple uses of the space. One is a human use, as the space of the disused building is temporarily occupied as a usable venue for an art exhibition. The other is a less human use as parts of the building itself become more or less suitable for the display of contemporary art. So, we see in some of the photographs that wooden scaffold inserts in the main bar space of the public house create a space for a number of hanging works. On the top floor, the only largely open space within the building becomes a space for performance. This radically alters what the building means, at least insofar as that meaning is manifested in the archaeological archive. Ordinarily, a building is built to be a certain thing and continues in that and other known uses before it becomes de-occupied and disused. At this point archaeologists arrive and record the building as a former example of whatever it was prior to its alteration. There is little space here for something so radical as temporary use as an exhibition space, even in a case such as this where remains of the exhibition were observed at the point of archaeological intervention. However, we see in these photos a phase of the building's use that has nothing to do with its previous life as a pub or with its later dual lives as university building and archaeological archive – recording the temporary unofficial use of moth-balled buildings.

*

The four sets of photographs of the same small group of buildings constitute a kind of 'sliding scale' of formality. Those taken by the professional archaeological photographer are technically accomplished: well lit, pared back and empty of human life, taken specifically to form part of an archive with an interpretive artificiality that creates one composite series of the building type from images of different units. My archaeological photographs are similar, but are much more informal aide-memoires rather than detached archives, taken in the moment to capture as much of an idea of the overall space as possible rather than just recording the physical structure, and are also much more numerous – almost 500 as compared with 64 – since they are designed for immediate reference rather than long-term archiving. The photographs by the contractor and the artist, meanwhile, are even less formal; poorly exposed and lit (where the flash can even serve to obscure material detail) and haphazardly composed, they act as personal record shots relating to specific activities. This scale, then, is not just one of formality but also of the record of human life in the space as distinct from the material architectural remnants – a distinction between a record of contemporary archaeology of lived space on the one hand, and of the remaining materials of a historic building understood as a type on the other.

Taken as a whole, these different photographic forms lurch from representing architectural traces to documenting lived space, from material to human. Understood as archaeological images, they remind us of photography's ability, which it shares with archaeological building recording, to do more than record remnants or ruins in purely nonhuman form – but to document human life over time through architectural space. Such informal photography, the documentation of human life rather than just material traces, could be included more often in archaeological site archives. And photography could be thought of, in the same way, not so much as a trace or a fragment of frozen time like a brick or a window frame, but an ongoing living record.

References

Bailey, D. 1990. 'The Living House: Legitimating Continuity'. In R. Samson (ed.) *The Social Archaeology of Houses*. Edinburgh: Edinburgh University Press, pp. 19–48.

Bailey, D. and L. McFadyen, 2010. 'Built Objects'. D. Hicks and M.C. Beaudry (eds) *The Oxford Handbook of Material Culture Studies*, Oxford: Oxford University Press, pp. 562–587.

Buchli, V. 2013. *An Anthropology of Architecture*. London: Bloomsbury.

Cairns, S. and J. Jacobs 2014. *Buildings Must Die: A Perverse View of Architecture*. Cambridge, MA: MIT Press.

Cole, S. 2017. *Photographing Historic Buildings*. Swindon: Historic England.

Dwyer, E. 2009. Underneath the Arches: The Afterlife of a Railway Viaduct. In A. Horning and M. Palmer (eds) *Crossing Paths or Sharing Tracks? Future Directions in the Archaeological Study of Post-1550 Britain and Ireland*. Woodbridge: Boydell and Brewer (Society for Post-medieval Archaeology Monograph 5), pp. 351–354.

Dwyer, E. 2015. *The Impact of the Railways in the East End 1835–2010: Historical Archaeology from the London Overground East London Line*. London: Museum of London Archaeology (Museum of London Archaeology Monograph Series 52).

Edensor, T. 2011. Entangled Agencies, Material Networks and Repair in a Building Assemblage: The Mutable Stone of St Ann's Church, Manchester. *Transactions of the Institute of British Geographers* 36(2): 238–252.

El Guinidi, F. 2015. Visual Anthropology. In H.R. Bernard and C.C. Gravlee (eds) *Handbook of Methods in Cultural Anthropology (second edition)*. Lanham, MD: Rowman and Littlefield, pp. 439–464.

Ellis, W. 2015. The Sea View Children's Hospital. https://abandonednyc.
    com/2015/05/20/the-sea-view-childrens-hospital/
English Heritage. 2006. *Understanding Historic Buildings: A Guide to Good Recording
    Practice*. London: English Heritage.
Garrett, B. 2013. *Explore Everything: Place-Hacking the City.* London: Verso.
Green, A. and J. Dixon 2016. Standing Buildings and Built Heritage. *Post-Medieval
    Archaeology* 50(1): 121–133.
Hicks, D. and A. Horning 2006. Historical Archaeology and Buildings. In D. Hicks and
    M.C. Beaudry (eds) *The Cambridge Companion to Historical Archaeology*.
    Cambridge: Cambridge University Press, pp. 273–292.
Historic England. 2016. *Understanding Historic Buildings: A Guide to Good Recording
    Practice*. London; English Heritage.
Johnson, M. 1993. *Housing Culture: Traditional Architecture in an English Landscape*.
    Washington, DC: Smithsonian Institution Press.
Pétursdóttir, Þ. And B. Olsen 2014. Imagining Modern Decay: The Aesthetics of Ruin
    Photography. *Journal of Contemporary Archaeology* 1(1): 7–56.
Salvadori, M. 1991. *Why Buildings Stand Up: The Strength of Architecture*. London:
    W.W. Norden.

9
PHOTOGRAPHING GRAFFITI

Alex Hale and Iain Anderson

Rarely solicited, and even less rarely welcomed in retrospect, there are fewer aspects of our built historic environment more contentious than graffiti. The occurrence of graffiti more often than not carries with it well-rehearsed attitudes towards both the art and the artist from the majority of the public and heritage sector alike. We seek to challenge this stance through unsettling traditional approaches to the use of photography in the archaeological recording of the built historic environment.

Our chapter focuses on two contrasting case studies in Scotland – Scalan, which is a nineteenth-century farm and Catholic seminary in the Cairngorms, and Pollphail, which is a 1970s oil workers village on the west coast of Argyll. We take stock of past recording techniques employed by archaeologists at these sites, in order to explore how new approaches to archaeological photography might be developed. The case studies form part of ongoing efforts to address graffiti art as a largely unrecognized form of cultural heritage in Scotland (Historic Environment Scotland 2016a).

Scalan has been a working farm since at least the 1700s. In two of the farm buildings are a dense concentration of written graffiti that document the life of the farm between the 1890s and the 1950s. The traditional history of Scalan is largely told through the life of the Catholic seminary, whereas the photographic recording of graffiti provides alternative narratives of working life on the farm, personal stories and harsh winters, to name a few. The second case study traces the place of photography in how six artists from the collective Agents of Change, transformed Pollphail through their complex, haunting works of art – through which the village came, until its demolition in late 2016, to be featured as part of the Argyll tourist board, 'Secret Coast Road', becoming a mecca for 'ruin'

* The authors would like to thank Philippa Elliot, Agents of Change, Steve Wallace and Zoe Ballantine (Historic Environment Scotland photographers), Neil Gregory (Survey and Recording Operational Manager at Historic Environment Scotland), Nick and Pip Molnar, John Toovey and Jennifer Stewart.

explorers. Our interest in this case is not just in the use of archaeological photography for recording, but in the inherently photographic affinities of graffiti through which the circulation of digital images shifted the village's significance within Scotland's historic environment.

Through these two examples, the chapter aims to make a contribution not simply to thinking about photography as a technique for recording ephemeral remains from the recent and contemporary past, but to examine how photographic images and archaeological sites can be intimately connected, shaping each other both through professional documentation and vernacular visual practices, in which digital photography today holds something in common with the inscription of buildings with graffiti in the past.

*

For Historic Environment Scotland, Scotland's national heritage body, to consider the photographic recording of graffiti within the National Record of the Historic Environment (NRHE) in anything but a happenstance manner, is new and in some cases unsettling. When considered alongside the broad range of survey work that the organization could be dedicating resources towards, graffiti is not seen by many as an obvious priority. But at a time in which definitions of Scottish heritage are being rethought (Historic Environment Scotland 2016b), we want to explore how graffiti can be understood as both a rich archaeological subject and historically significant artefact in its own right (Giles and Giles 2010; Lovata and Olton 2015; Merrill 2015), and therefore as an important target for photographic recording.

The use of photography to record graffiti archaeologically is, of course, not new (Cooper and Chalfant 1984) – but its recording from the very recent and contemporary past has been less common. Previously the scope of the graffiti that has been regarded as being of interest has often been limited, within an incredibly broad corpus for potential recording. Viking runes scratched on to the Neolithic tomb of Maes Howe on Orkney have been classified as graffiti, and therefore form part of the same thematic grouping as a tag sprayed onto a wall yesterday (Forster *et al.* 2012). From more recent periods, military graffiti from the twentieth century has often been recorded (Cocroft *et al.* 2006), and the recording of 1970s Sex Pistols graffiti in Denmark Street in Soho, London (Graves-Brown and Schofield 2011) has already disrupted and questioned the temporal limits of the archaeological recording of graffiti.

Scalan (Figure 9.1) is a traditional subject for archaeological survey. Located in Moray, it comprises a group of listed historic, rubble stone farm buildings dating from the late nineteenth and early twentieth century (with earlier parts incorporated). The farmstead includes two mill buildings: the older, L-plan north mill, and the rectangular plan south mill. The north mill houses a cart-shed with first floor haylofts, threshing floor and well-preserved, intact threshing machinery driven by

Figure 9.1 Archaeological photographer Zoe Ballantine recording graffiti at the north mill at Scalan, Braes of Glenlivet, Scotland, 2016. During a two-day recording visit the survey team recorded seventy-four separate locations of graffiti in the two mills. The earliest date places the graffiti written in 1874 (Historic Environment Scotland DP246747).

a waterwheel contained in a wheelhouse on the west side. The off-shoot of the L-plan retains a stable with timber horse stalls. The south mill, at its south end, retains the remains of an external undershot waterwheel, fed by a timber lade, which drove an internal threshing machine. The lower, north end of the range remains fitted out with timber byre stalls and a cobbled floor. The buildings are part of the Catholic seminary that was established here in the early part of the eighteenth century. Found throughout both mill buildings are some of the finest and varied examples of everyday graffiti in Scotland, dating from the late-nineteenth century through to the mills' period of decline in the late-twentieth – remains that are similar to those discussed in detail by Mel Giles and Kate Giles based on the rural barns of the Yorkshire Wolds, where graffiti by 'horse lads' and wartime 'land girls' represents a unique document of social history, including the everyday lives of women in the twentieth century (Giles and Giles 2010). Like many rural buildings of this type, the mills at Scalan appear to have slowly fallen out of use and simply had their doors closed, with no development pressures or residual building value to bring threat of alteration or demolition. They have been part of the Scalan Association, which was primarily set up to maintain the Catholic seminary building and heritage, and who have paid for the upkeep of the buildings and prevented the mills from falling into disrepair. Thus, the mill buildings sat quietly accumulating heritage significance as little-altered examples of their type, a value recognized by listed building status in 1972 and further in the present day, through heritage funding to improve access and understanding for visitors.

The graffiti within the mills ranges from the late nineteenth century through to the 1950s, spanning much of the useful lives of the buildings. In terms of the composition of the graffiti, the locations across both mills are worth considering. In the north mill there are two areas of dense concentration of writing: one in the area surrounding the threshing machinery, including on the wooden panelling of the machine itself, which was purpose-built into the mill, and a small room adjacent to the machinery that was used to fill sacks of threshed grain (Figure 9.2).

Figure 9.2 The sack room in the north mill at Scalan, with an inset showing detail of the graffiti (photographed 2016). The room would have provided a warm, but dusty environment, to await the threshed grain to pour out of the hopper. There is a dense concentration of graffiti in this space, reflecting a sustained time during which the farm workers spent in this space (Historic Environment Scotland DP242908 and detail DP242919).

Another area of dense graffiti is in the loft above the threshing machine. In the south mill, one of the concentrations is on the panels of a door that separated the barn from the byre and in the hayloft at the south end of the building. The graffiti is concentrated in these particular areas because these were the places where the mill workers were stationed before, during and after threshing the grain: they would stand waiting for the crop to be brought into the mill, would winnow it in the barn, pass it through the threshing machine and place sacks at the mouth of the hopper to collect the grain.

A full measured buildings survey and extensive photographic record of Scalan mills was made in 2013, by the Royal Commission on Ancient and Historical Monuments of Scotland – but this survey only came to include one image that was specifically designed to record some of the graffiti found in the buildings and the majority was omitted from the survey, despite its significance as an element of the historic buildings. The density of written graffiti across both mills provides primary visual evidence of everyday activities and domestic lives that played out on-site: quantities of harvests and livestock are documented in a range of locations, which can be used to interpret the productivity fluctuations of the farm, leading to understanding the micro-economics of this part of Scotland, which could be tied into more regional or national trends. Specific examples of everyday farm activities are recounted in some detail, such as how much fencing was required to enclose the farm, who undertook this role and how much the materials cost in the 1920s. On a range of locations are notes about the weather and one particular door records details for specific periods of the year between 1909 and 1930 (Figure 9.3). As well as the everyday, there are the global historical events that affected the farm. For example, various members of the farm took part in the world wars and aspects of their departures and returns are documented in the graffiti.

*

Unlike Scalan, Pollphail's graffiti offers different challenges – a disruption in the site history rather than history intertwined (Figure 9.4). The worker's village at Pollphail became something of a phenomenon within certain circles of Scottish culture during its final years. The redevelopment bulldozers finally moved in, during late 2016, bringing to an end one of the most unusual built histories of recent times. Pollphail, situated on the west coast of the Cowal peninsula in Argyll, was chosen as the site for an oil platform production yard in the mid-1970s, and the village was built between 1975 and 1977 (Walker 2000). Its location on Loch Fyne gave access to deep, coastal waters, and a large geological basin was taken advantage of for the site of the production yard. A few hundred metres to the south a large complex to accommodate the necessary workforce on the otherwise sparsely inhabited peninsula, was built. The workers' village comprised a multi-functional centre that housed facilities such as reception,

Figure 9.3 Columns of weather reports hand-written onto the mill wall at Scalan, circa 1905–1920 (photographed 2016) (Historic Environment Scotland DP242983).

laundrette, TV lounges, function suite, bar, kitchens and catering. Terraced along the hillside below were rows of single-bedroom apartment blocks and to the north of the centre, small terraced houses were built for management. It was designed by Thomas Smith, Gibb and Pate Architects and has been described more recently as being 'sadder than a deserted holiday camp' and comprising 'monopitch roofs covered in profiled metal sheeting, rendered rounded stair towers and walls clad in horizontal timber boarding' (Walker 2000, 432).

Once built, the story of Pollphail takes its first twist, with the scrapping of the plans for the production yard and the abandonment of the workers' village site at a late stage. The kitchen and laundrette were fitted out with machinery, there was furniture in the dormitories and room keys at reception. The site was abandoned before it had been occupied and sat silently in that condition for a number of

Figure 9.4 Graffitied gable ends of the accommodation block that greeted visitors at the 'ghost village' of Pollphail, Argyll, Scotland, 2016. The collaborative piece combines a range of graffiti styles, from blow-up to street art mural. Since being painted in 2009 the fresh colours stood in contrast against the concrete render and active ruination of the village (Historic Environment Scotland DP246442).

years, too far from large settlement to attract significant vandalism and providing a spectacular playground for the few local children. There were occasional plans for alternative uses, but nothing that ever came to fruition and as time passed, the habitation of the complex became a more unlikely proposition. Development plans at the neighbouring deep water dock overtook the workers' village. By the early 2000s a boating marina was established nearby, and in 2013 plans were approved for a major tourist destination and the expanded marina that is active today.

As the marina development took off, knowledge of the workers' village began to spread for a very different reason. It became a regular attraction for artists and urban explorers, who regularly published images of the site on various blogs, forums and photo sharing sites. For example, local photographer Philippa Elliott's images (Figure 9.5) captured beautifully the suspension of time and sudden abandonment that characterized the site and these were widely shared around the Internet, spreading images of Scotland's 'ghost village' (Urban Glasgow 2009).

The burgeoning interest of the artistic community in Pollphail appears to have been embraced by the site owner, and in parallel to creating new redevelopment plans for the complex, they made the unusual decision in the summer of 2009 to

Figure 9.5 Keys hanging on dormitory hooks at Pollphail in 2015 (photo: Philippa Elliot).

allow an international collective of artists, Agents of Change to use the village complex as a canvas for their work. This collective uses graffiti art as a medium to, in their own words, 'attack space' (Agents of Change 2018). Their work is globally located and appreciated; featuring in situations as diverse as gallery exhibitions through to car branding, as well as a huge and expanding corpus of street art. Working at Pollphail offered the collective an opportunity to collaborate on a scale that is rarely possible. Six artists – Derm, RemiRough, Juice 126, Jason System, Timid and Stormie Mills – worked on site for three days, producing artworks. Their 'attack space' approach involved visiting, painting intensely, and leaving, with little initial publicity or explanation. Their work was left for others to discover, allowing word of mouth, social media and the aforementioned blogs and forums to disseminate the news of their intervention. Given the nature of the village and the fabric of the buildings, Agents of Change had a broad range of canvasses to choose from. Their overall aims are clearly articulated on their blog:

> Working together on huge collaborative walls and individually in hidden nooks and crannies all over the site the artists realised long held dreams and were inspired by the bleakness and remoteness of the site. Drawing on the history of the village the artists' stated intent on completion of the project was to populate the Ghost village with the art and characters that it deserved.
>
> Agents of Change 2018

For Historic Environment Scotland, at Pollphail, there was, unlike at Scalan, no pre-defined 'heritage asset'. However, the artists themselves made active use of the geography and built environment of the village. Visitors were met by the neighbouring gable ends of two terraces, upon which a collaborative mural featured the work of all six artists, the paint-staining, tags and swatches of which can be seen in Figure 9.4.

One particular graffiti piece painted by Jason System (Figure 9.6) comprised a profile portrait of Juice 126 with his mouth open, and out of his mouth a speech bubble emerges with two lines of numbers: '55.870056 –5.306956' are the latitude and longitude location of Pollphail and when positioned through online mapping tools, locate the viewer on the central recreation block of the village. The work puts Pollphail on a map through digital media by enabling the 'remote' visitor, when viewing a photograph of it, to discover its location – creating new visual geographies for artworks that parallel more recent attempts by museums and galleries to engage with location-based dimensions of social media, as for example with the Tate's Artmaps project, which enabled participants to identify the locations of the paintings that the Tate holds in its collection (Tate 2018). In the case of the art at Pollphail, System has disrupted this approach by omitting both gallery location and the need for his graffiti piece to have an extended physical lifespan, which might lead us to muse upon the role of the digital archive within this artistic transaction.

The work of Stormie Mills at Pollphail also reads and engages with the built environment in a distinctive way. He is renowned globally for his iconic character

Figure 9.6 Graffiti by System showing Juice 126 with latitude/longitude grid coordinates (55.870056–5.306956) in a speech bubble, Pollphail 2009 (photographed 2016) (Historic Environment Scotland DP24644).

Figure 9.7 Two of Stormie Mills's characters moving towards the central courtyard at Pollphail. They are situated on the peeling render of a stair tower, adjacent to a piece by Derm, in a recessed doorway (Historic Environment Scotland DP246443).

of a largely monochrome anthropomorphic form, situated within, on or against its surroundings. At Pollphail, Stormie Mills created 21 of his characters at a range of locations and in different forms (Figure 9.7), each moving towards the open courtyard space on the east side of the main communal block. Perhaps some of the most striking pieces are those that appear on the exterior walls of the accommodation blocks – one points at its stomach and appears to be moving towards the central refectory, another is a painted washing machine and clothes drier that has been turned into a face with dark eye sockets that are formed by the machine doors (Figure 9.8). Stormie Mills and (Jason) System painted the two images by spray painting the utilities. The image of two faces staring out from the industrial laundry machines creates a juxtaposition between unused and abandoned, with functional machinery and artistic creation of haunted humanity. The expressions on the two faces comprise hooded ferocity on the left by System, with tearful acceptance on the right by Stormie Mills. The choice of

Figure 9.8 Washing machines and tumble driers painted by System and Stormie Mills at Pollphail, 2009 (photographed 2016). Timid's drip piece on the far left escaped recognition by the survey team until May 2016, prior to the demolition of the village in November 2016 (Historic Environment Scotland, DP242165).

spray can colours, blue for the left, creates a confusion of serenity with hooded form, an image so often appropriated as an emblem of 'disorder', as geographer Tim Cresswell has discussed (1991). It was only after our third visit to Pollphail that we recognized a further piece by Timid immediately to the left of the washing machines. His works of running paint down walls forming dripping pieces were previously unrecognized and had been confused with the 'natural' decay of the village. After 2009 a number of graffiti artists were attracted to Pollphail and they made contributions through further artistic interventions, much of which adds to and complements the graffiti by Agents of Change, but can also complicate the identification of their complete corpus of work.

*

As the field recording progressed at Pollphail and Scalan, we recognized that although two very different places, these sites provided the focus for us to explore the methodologies for photographing graffiti. The archaeological photographic survey started with general views of the complete site, followed by shots of individual buildings, then elevations of walls containing graffiti, and detailed images of individual pieces. The archaeological photographic approach departed from usual practice by bringing an archaeologist and a photographer into the field together, so that the location and detail of each photograph could be recorded. The graffiti at Scalan present an ambience of a specific period of the site's use, adding another layer onto the vernacular architecture. The specific locations within the buildings themselves provide the backdrop and the framing devices for all of the graffiti images created. High resolution, photographic images were taken of each panel, door jamb, rafter, stable stall, wall or timber fitting in both mills.

The medium in which the graffiti was written also influences the photographic methodology. The graffiti of Scalan is predominantly written by hand in pencil, with some crayon and a little chalk used. So the colour palette tends towards greys with some blue crayon and a little white and yellow chalk. But this predominance of graphite grey provided a suitable contrast against what would have been freshly sawn timbers of the internal fittings at the time of the graffiti's creation. Over sixty years on from the last dated graffiti we are faced with a weathered context, which dulls the clarity of the writing. Therefore the photographic recording was carefully considered to provide the clarity of images that would be required for later transcription or post-production research uses.

Similarly, the positioning of the writing does have a consideration behind much of it and although we are at an early stage of documenting each piece of writing, there are a number of points that can be drawn out. First, the quotidian writing often comprises quantities, dates and people. In many cases this form of graffiti is positioned on or at doorways, which lead between work areas. Second, personal stories are found in small groups some of which are relatively secluded and partially hidden from sight. Another facet that leads us to create a particular image of Scalan's past is the style of the writing and the words inscribed. The majority of the writing is in late Copperplate, cursive script that was taught in Highland schools at the time. Finally, apart from the writing there is a huge number of images of animals and people.

The images form a corpus of wild and domestic animals and a range of people who were obviously characters from the local area (Figure 9.2). By photographing each image we can build up a corpus of individuals and potentially identify them with local individuals. In addition, the animals range from weasels to horses, but the predominant image appears to be the pig, for a currently unknown reason. Some of the human figures comprise horned people, who perhaps were known to the illustrators by their behaviour towards them and others. The photographs from Scalan document graffiti from a period of huge social change, and offer a resource for a kind of visual history that complements oral and community history in its attention to specific lived experiences in particular locations.

Pollphail village was never going to be a candidate for protection as a monument and so the photographic documentation of the graffiti in the workers village at this final stage of existence was an exercise in 'preservation through record'. The graffiti intervention was a disruptive visual event in Pollphail's decline towards demolition, bringing a new, albeit temporary, and alternate value and function into the process of ruination, which was always known to be finite. The photographic archive involved archaeological acts of discovery and record that extended the existence of the demolished village, and also brings a new dimension to the visuality of the graffiti through the camera and the mapping of images. Many people visited and recorded Pollphail between 2009 and 2016. These visits have created many photos in the 'urbex' (urban exploration) style, which have a heritage of their own and exist alongside, but rarely ever mutually cross-reference or acknowledge 'official' visual archives of national heritage and archaeology. The vast majority of visitors had no intention of curating an accurate documentary record, and many of the artists themselves regard their graffiti art as interventions that react to the site in the moment of their production, and have little interest in retrospectively revisiting and discussing their work.

*

In both Scalan and Pollphail, the understanding and significance of heritage in both places is bound up, through graffiti as a visual medium, with photography. It is impossible to consider the value of the Pollphail site without reference to the digital social media through which it was partially created. At the two sites, the graffiti ranges from everyday records of nineteenth- and twentieth-century life to potentially unsettling, artistic responses to a place that can be loosely grouped as street art. What forms of graffiti lie in the middle-ground between the historic document of Scalan and the artistic intervention of Pollphail? Somewhere along this scale we must position and address the unloved and unwanted aspects of graffiti, such as the tags and hand-styles that developing artists are practising. These interventions often carry negative attitudes to such graffiti and come in the most part from its use as a form of off-the-cuff protest/territory marker/'boredom breaker', which goes against the conventions of 'polite' society and environment.

The democratization of heritage values is currently a matter of some debate across the wider sector of heritage studies (Jones and Leech 2015; Jones 2017) and breaking this down alongside the background, politics and culture of graffiti has been addressed elsewhere (Lovata and Olton 2015). However, the photographic dimensions of heritage valuation in the context of digital media needs to be addressed. Graffiti, as a site for the mutual photographies of archaeological recording and vernacular sharing – represents a key example here. As Colin Sterling has shown in his discussion of how 'clichéd' tourist photographs operate to shape monumental heritage (Sterling 2017), so in the context of graffiti the photographic image can become part of the material site, even after its demolition. Just as modern and contemporary heritage such as graffiti from the recent past is increasingly taken seriously, rather than treated as damaging buildings and landscapes, so too must non-professional forms of photographing and sharing images of the built environment, which are often discussed in pejorative ways as snapshots (Sterling 2017).

The opportunity to use the testimony of artists themselves, in the case of Pollphail, adds another dimension to questions of documentation, authenticity, value and cultural appropriation. The statutory protection of graffiti, preserving it as public art or monumentality, would in the case of contemporary graffiti destroy the central element of its character – its ephemerality. At the same time, its photographic dimensions offer, in a world of digital and social media, new opportunities for thinking about documentation – who takes the photograph, who selects, curates and shares images and knowledge, and how these actions can change places and give them value for communities.

Rather than just representing an instance of a largely unrecognized form of modern heritage that can be brought into the field, graffiti's relationship with photography leads us towards rethinking the distinctions between sites and their archives, and the material and the visual, in archaeology. The practice of photography can be in such cases a form of engagement and dialogue with communities, a two-way, reciprocal process in which photographs have the ability to reshape the immediate past. As practice of inscription that can alter the historic built environment, to take a photograph is not so unlike the making of graffiti – a similarity that adds another dimension to how we think about the relationships between Archaeology and Photography.

References

Agents of Change. 2018. Agents of Change website. http://agents-of-change.co.uk

Cocroft, W., D. Devlin, J. Schofield and J.C. Thomas 2006. *War Art, Murals and Graffiti: Military Life, Power and Subversion*. York: Council for British Archaeology Research Report 147.

Cooper, M. and H. Chalfant 1984. *Subway Art*. London: Thames and Hudson.

Cresswell, T. 1991. The Crucial 'Where' of Graffiti: A Geographical Analysis of Reactions to Graffiti in New York. *Society and Space* 10: 329–244.

Forster, A., S. Vettese-Forster and J. Borland 2012. Evaluating the Cultural Significance of Historic Graffiti. *Structural Survey* 30(1): 43–64.

Giles, K. and M. Giles 2010. Signs of the Times: Nineteenth – Twentieth Century Graffiti in the Farms of the Yorkshire Wolds. In J. Oliver and T. Neal (eds) *Wild Signs: Graffiti in Archaeology and History*. Oxford: Archaeopress (Studies in Contemporary and Historical Archaeology 6), pp. 47–59.

Graves-Brown, P. and J. Schofield 2011. The Filth and the Fury: 6 Denmark Street (London) and the Sex Pistols. *Antiquity* 85: 1385–1401.

Historic Environment Scotland 2016a. *Investigating and Recording Scotland's Graffiti Art (Phase 1) Project Report*. Edinburgh: Historic Environment Scotland.

Historic Environment Scotland 2016b. *Scotland's Archaeology Strategy*. Edinburgh: Historic Environment Scotland. http://archaeologystrategy.scot/files/2016/08/

Scotlands_Archaeology_Strategy_Aug2016.pdf

Jones, S. and S. Leech 2015. *Valuing the Historic Environment: A Critical Review of Existing Approaches to Social Value*. Manchester: AHRC Cultural Value Report https://www.escholar.manchester.ac.uk/api/datastream?publicationPid=uk-ac-man-scw:281849&datastreamId=FULL-TEXT.PDF

Jones, S. 2017. Wrestling with the Social Value of Heritage: Problems, Dilemmas and Opportunities. *Journal of Community Archaeology and Heritage* 4(1): 21–37.

Lovata, T. and E. Olton (eds) 2015. *Understanding Graffiti: Multidisciplinary Studies from Prehistory to the Present*. Walnut Creek, CA: Left Coast Press.

Merrill, S. 2011. Graffiti as Heritage Places: Vandalism as Cultural Significance or Conservation Sacrilege? *Time and Mind* 4(1): 59–75.

Merrill, S. 2015. Keeping it real? Subcultural graffiti, Street Art, Heritage and Authenticity. *International Journal of Heritage Studies* 21(4): 369–389.

Oliver, J. and T. Neal (eds) 2010. *Wild Signs: Graffiti in Archaeology and History*. Oxford: Archaeopress (Studies in Contemporary and Historical Archaeology 6),

Stationery Office 2014. *Historic Environment Scotland Act 2014*. Norwich: Stationery Office http://www.legislation.gov.uk/asp/2014/19/contents/enacted

Sterling, C. 2017. Mundane Myths: Heritage and the Politics of the Photographic Cliché. *Public Archaeology* 15(2–3): 87–112.

Tate 2018. Artmaps website. http://artmaps.tate.org.uk/artmaps/tate/#zoom=15&lat=51.51&lng=-0.10&maptype=hybrid

Urban Glasgow 2009. Polphail, the Ghost Village at Portavadie, Argyll. http://urbanglasgow.co.uk/archive/polphail-the-ghost-village-at-portavadie-argyll__o_t__t_1500.html

Walker, F.A. 2000. *Argyll and Bute*. London: Penguin (Pevsner Architectural Guides: Buildings of Scotland).

10

PHOTOGRAPHY, ARCHAEOLOGY AND VISUAL REPATRIATION

Samuel Derbyshire

This chapter considers the potential value that historic ethnographic photographic collections might bring to archaeological engagements with the recent past. It does so via a discussion of recent research conducted in the Turkana region of northern Kenya between 2014 and 2015. Much of this fieldwork took place around a rich photographic collection, comprising images taken by numerous individuals who journeyed to Turkana at different times over the last century. I first encountered this collection in 2013 at the Pitt Rivers Museum, Oxford, and brought it to Turkana a year later having supplemented it with my own photographs of the Turkana-related ethnographic object collections both at the Pitt Rivers Museum and at a range of other institutions. As I outline below, my use of these photographs was deeply informed not only by recent developments in visual anthropology and photo elicitation, but also by approaches to the recent past that have emerged from the worlds of postcolonial African archaeology and indigenous archaeology. During discussion sessions and interviews, the collection I brought to Turkana served to evoke a series of historical narratives that contrast

* I wish to thank Lucas and Eliza Lowasa, and the Lowasa family of Nakurio, Turkana, for their advice, support and kindness, and the editors of this volume for their invitation to contribute. I would also like to thank the Pitt Rivers Museum, Oxford, for my extensive use of their photographic archives, and especially Chris Morton for the valuable insight given before, during and after the research was undertaken, and for assistance working with the photographic archives. The research for this article was carried out under the National Commission for Science Technology and Innovation (NACOSTI) permit no. 14/5212/833. I wish to thank the Royal Geographical Society, the Frederick Soddy Trust, the Clarendon Fund – University of Oxford, the British Institute in Eastern Africa and the Turkana Basin Institute for financial and logistical support.

with and disrupt still-prevalent notions of the region's contemporary population as socially static and passive in the face of broad economic, political and environmental change. Before exploring the fieldwork in more detail, I want to first consider its theoretical foothold in the interplay between several distinct bodies of literature and outline some of the ways in which both archaeology and photography have been actively implicated in the production of such prevalent visions over the last century, and the ways in which these visions are politically consequential.

The last few decades have seen a monumental shift in the ways in which museums containing historical ethnographic collections are envisioned, engaged with and utilized in the contemporary world. Laura Peers and Alison Brown have described this development as 'one of the most important in the history of museums' (Peers and Brown 2003: 1). Often uncomfortable bodies of everyday objects, photographs, films and other materials have ascended to an unlikely position at the vanguard of much anthropological research. This has taken place not in denial but rather direct confrontation of anthropological museums' earlier entanglements with the colonial project, their complicity in the production and representation to Western audiences of visions of primitiveness and otherness and their accompanying roles, in conjunction with early ethnographer partners and contributors, in disseminating and incubating 'literary and theoretical savages' (MacDougal 1997: 279) in popular culture.

This troubled heritage and its enduring legacies in more recent displays, representations and outreach/partnership policies was the subject of much critique in the 1990s (Clifford 1991, 1999; Coombes 1994; Harlan 1995; Barringer and Flyn 1998). From these debates, James Clifford's conceptualization of museums as 'contact zones' – spaces where different cultures, worldviews, interests and stakeholders encounter each other and contest the past in an ongoing 'historical, political, [and] moral relationship', has perhaps been the most influential (Clifford 1999: 192–3). The point that these meetings and conversations are ongoing, unfinished and open to the incorporation of heterogeneous pasts and multi-vocal accounts has become critical in numerous research and outreach projects that have taken part in transforming the one-way relationship that had previously existed between museums and their source communities (Peers and Brown 2003; Marstine 2006; Crooke 2008). Museum collections are no longer seen as dead places, lifeless vestiges of colonialism and imperialism, but rather opportunities to open up new ways of thinking and talking about both past and present – tools that might help to subvert the colonial gaze (and its legacies) and find new values, histories and meanings in its remains.

In no other museum-based medium is this open-endedness and lack of fixity more profound and valuable than in historical ethnographic photographs; they are, as Elizabeth Edwards has written, 'the most potent of museum objects' (2003: 83). Their simultaneous qualities of stillness and restless mutability, the traits that often made them such persuasive indicators of the timelessness and pristineness of societies living at the margins of empire, also imbue them with the powerful potential to express counter narratives that work against the gaze of the camera lens, and to articulate a diversity of new versions of the times, histories and transformations that they depict (Poignant 1994; Edwards 2001, 2003). The place of photographs in anthropological museum collections has, in many cases, been unveiled as inconclusive and temporary, a singular stage in a still-unfolding biography (Bell 2003). Visual repatriation, and the often-accompanying field method of 'photo elicitation' (Collier and Collier 1986), have seen these objects take part in, and become tools that are used to tell, neglected and supressed histories in the locales that they pertain to – rather than simply remaining as enclosed visions of the past that are read in silence and obscurity.

Whilst the development of the practice in anthropology of returning such photographs to source communities and using them to seek to create new knowledge is not without ethical implications (Niessen 1991; Thomas 1997), it has nevertheless provided many museums with a much-required means of building relationships with source communities. In the process, many communities across the globe have reintegrated collections of historical photographs into local knowledge systems, inscribing personal experiences of past places, people, events and objects on to them and using them to address ongoing social, economic and political issues (Binney and Chaplin 2003; Bell 2003, 2008, 2010; Carrier and Quaintance 2012).

Archaeology, meanwhile, has been undergoing a parallel and similar transformation. In moving away from previous implicit correlations between history and European encounter or occupation (and thus textual records), postcolonial historical archaeologies have sought to envision history instead as a kind of community memory transmitted across generations in numerous conscious and unconscious ways (cf. Lane 2004; Pikirayi 2004; Schmidt and Walz 2007; Davies and Moore 2016; Straight *et al.* 2016). As part of this shift, archaeological research undertaken in postcolonial contexts has become less about merely collecting evidence of the past for interpretation and presentation elsewhere, and more about new forms of collaboration with communities (Greer *et al.* 2002; Harrison and Williamson 2004; Schmidt 2010, 2014; Davies *et al.* 2014; Schmidt and Pikirayi 2016) with the aim of co-creating new indigenous historical narratives that are valuable or 'useable' in the contemporary world and connected local and global political issues (Lane 2011). In Africa, recent years have thus seen archaeology used as a means of engaging with post-genocide development in Rwanda (Giblin 2010, 2012, 2014), 'supermodernity', conflict and destruction in Ethiopia (Gonzáles-Ruibal 2006, 2008) and failed development interventions in Kenya (Davies and Moore 2016; Derbyshire 2018; Derbyshire and Lowasa forthcoming). In such contexts, it is archaeology's predilection for quotidian artefacts – the 'materialities of everyday life' (Schmidt and Walz 2007:

142) and 'the small things forgotten' (Deetz 1977) – that renders it uniquely capable of seeking out and documenting 'the voices of the subaltern, the Other, those who have no voice in official records' (Gonzáles-Ruibal 2008: 248; but see Spivak 1988).

This divergence and yet mutual significance of postcolonial historical archaeologies and recent developments in visual and museum anthropology struck me when I was surveying the extensive ethnographic objects and photographs from Turkana in the archives of the Pitt Rivers Museum in 2013. Both enterprises seemed to me to be grappling with the same objectives, confronting similar troubled disciplinary pasts and finding new ways of inserting themselves into non-Western epistemologies and narratives that the broader disciplines of archaeology and anthropology could, in turn, be influenced by and perhaps even re-imagined through. In doing so, however, they were forging new methodological paths that drew close but did not fully converge.

*

Numerous features of the collection from Turkana advocate its relevance for archaeological analysis and interpretation. Albums of ethnographic photographs and accompanying artefacts are, after all, the deposits made by various interventions into Turkana through the years, 'the cumulative remains of multiple past processes' (Lucas 2004: 38) existing as a kind of palimpsest many miles away from the initial sphere of their enactment. Indeed, the Pitt Rivers Museum contains a uniquely wide collection from Turkana, the earliest objects and images dating from the earliest years of colonial rule at the dawn of the twentieth century, and their latest donated in the 1990s. Taken together, these layered residues document gradual changes in the production, use and exchange of a range of everyday objects throughout the various social, economic and political transformations that have manifested themselves in northern Kenya over the last century. As is the case in numerous archaeological sites, this has largely been the result of accident rather than purposeful preservation. The changing materiality of everyday life is thus not necessarily patently discernible in the collection, but can rather be located in backgrounds and subtle details – the visual presence or absence of shells in female neck ornamentation, for example, or the shift from leopard and baboon skins to large cotton blankets in male clothing.

It also became apparent to me during my early encounters with the collection that by accidentally apprehending these details it had served to undermine the vision of Turkana that many of the individual albums and photographs were initially designed to portray. Images of a romantic 'vanishing race' had, over time, accumulated into a much larger and more detailed picture of a dynamically changing population, a cluster of communities who have continuously incorporated new materials into a changing socio-material universe, reimagining various practices, institutions and objects along the way. The colonial gaze had been undermined and decentred by the passage of time, by the very materiality of the 'colonial visual economy' and its physical endurance beneath a slew of later deposits. To a place like Turkana – still beleaguered by the legacies of its colonial past and regularly represented through various visual tropes in the popular press as a timeless traditional world, the home of a culture condemned to annihilation by globalization and the trappings of modernity (cf. McCabe 2004; Lane 2015) – the counter narratives captured in the Pitt Rivers Museum collection could not be more significant or more urgent.

It was the recognition, however, that these narratives (which I could only catch quiet and vague glimpses of) were dormant and unable to surface away from the memories, knowledge, experiences and voices that might fully disclose them that led me to formulate a visual repatriation project. I initially imagined that such a project could perhaps 'convey a past that had not died in individual memories, but which had been suppressed in the European-recorded historiography' (Binney and Chaplin 2003: 100), as was the case with the early twentieth-century collection of photographs of Maori elders visually repatriated by Binney and Chaplin. However, unlike this and many other like-minded projects, I wanted to find out what might emerge from a fuller temporal range of photographs and a focus on changing material culture. It was with these intentions that I amalgamated an assortment of historical ethnographic photograph collections into a single chronicle, which I augmented with my own photographs of various Turkana artefacts from the Pitt Rivers Museum and other collections and brought to southern Turkana in June 2014. Elizabeth Edwards (2003: 90) maintains that people do not merely talk about historical photographs but rather 'talk with them and weep over them', their ambiguity and silence serving to transform them into tools that can be used to tell numerous divergent accounts of the past. I wondered what they might do as archaeological tools. What would an archaeology of Turkana's recent past, performed by members of its contemporary population, look like?

Such a question naturally invokes the emerging theories and methods of 'indigenous archaeology', a multi-seeded and multiform sub-discipline orientated towards research that is undertaken 'by, for and with Indigenous communities' (Colwell-Chanthaphonh et al. 2010: 228; Silliman 2008). Indigenous archaeology has grown to maturity in Australia (Harrison and Williamson 2004), New Zealand (Rika-Heke 2010), North America (Watkins 2000, 2005) and Canada (Nicholas and Andrews 1997; Nicholas 2010) where there are currently numerous innovative projects and lively ongoing debates. However, in sub-Saharan Africa, despite the development and expansion of postcolonial historical archaeologies in more recent years, it has not yet convincingly taken root (Lane 2011). This is not to suggest that there is a dearth of archaeological research that involves close collaboration and involvement with various local communities (see Chirikure *et al.* 2008; Denbow *et al.* 2009; Davies *et al.* 2014; Schmidt and Pikirayi 2016). Rather, these projects have chosen not to explicitly designate their approaches as 'indigenous archaeology' for a number of reasons. One such reason is perhaps the fear of inadvertently essentializing the cultures and practices of participating communities by structuring research activities around the term 'indigenous'. This term, as Lane (2011) points out, does not hold the same significance in African contexts as it does in countries and continents where self-identifying Indigenous groups are often marginalized, and indigeneity correlates with a sense of firstness or nativeness in relation to the descendant populations of settlers (South Africa is a notable exception to this; cf. Yeh 2007).

Nevertheless, the objectives, methods and practices of many postcolonial archaeological research projects that involve close collaboration with local populations in Africa converge strongly with those of indigenous archaeologies elsewhere. The research I outline in this article might be seen as a methodological contribution to both spheres, in the sense that it describes an archaeological use of historical photographs for the elicitation and documentation of historical narratives rooted in local ontologies and epistemologies rather than Western-rooted accounts of time and associated conceptualizations of social change and history. Having said this, I do not presume that conducting the fieldwork necessarily brought any tangible benefits to the various communities and individuals who participated along the way (which is often a primary objective of indigenous archaeology) aside from the opportunities it afforded people to engage in some vigorous conversations and to observe and discuss resources that are not locally available. Moreover, I must concede that both the strategy I deployed in the field and the selection and organization of the photographic collection that was visually repatriated were my own doing, rather than that of the various communities who participated in the research. Both of these factors perhaps correlate with the broader point that the primary significance of this research is its facility to produce narratives that may contest enduring and widely held problematic suppositions regarding social change and historicity in Turkana which, as I outline below, still influence development planning and practice, rather than its applicability or engagement with immediate local needs or aspirations. This is not to say, however, that the former might not benefit the latter in the longer term.

*

It is a country of barren and immense aspects, twisted and rendered by the cataclysmic convulsions of a long-ago past.

BROWN 1989: 16

Before outlining some of the discussions that took place around the photographs on their return to Turkana, and the various historical narratives and arguments that emerged, it is worthwhile to consider the ways in which both archaeology and photography have contributed to Turkana's 'dehistoricization'. This, I argue, has taken place through their entanglement and representation of particular forms and qualities of time and corresponding neglect of others. Monty Brown's account of Turkana from his work *Where Giants Trod* not only conveys its profound emptiness and openness, but also the persistent entanglement of these features with notions of timelessness and the enduring touch of a 'long-ago past'. Such descriptions of Turkana are common, and may be found in much of the popular literature relating to the region. Its international renown as the location of some of the oldest human fossils ever discovered by palaeoanthropologists and archaeologists (Leakey and Lewin 1979; Wood and Leakey 2011), and more recently the oldest known stone tools ever made and used by hominins (Harmand *et al.* 2015), stand in stark contrast to the pervasive indifference and disregard with which its more recent history is habitually addressed. A strong case might be made for the economic and political advantages to be wrought by local communities through the enduring prominence of Turkana's deeper history, and the general emphasis that is placed on the region's significance as an origin point for all human life. However, the local benefits of this dominant and overarching past have arguably so far paled in comparison to the roles it has played in obscuring the history and heritage of Turkana's contemporary population.

Whilst the general obfuscation of this more recent past is without doubt negative in its own right (in terms of academic understandings of the history of the region), perhaps even more troubling are the social, economic and political ramifications of this process. Turkana's prominent place at the beginning of a globally orientated story of human evolution and expansion (a story that evokes a sense of belonging in many foreign visitors to the region) seems to have coalesced with a range of tendencies and proclivities in broader understandings of rural non-industrialized (and particularly pastoralist) African societies. These co-functioning circumstances have served to silence local accounts of time and change, in many cases suppressing the claims of autochthony and belonging that they articulate and the visions of the unfolding future that they disclose.

The fraught relationship between Turkana's deep past and the history of its contemporary population is perhaps symptomatic of a deeper tension between the different forms of time that are currently at play within its boundaries. Indeed, the dichotomy of measured time versus experienced time, what Gosden (1994: 2) refers to as 'a basic antinomy in views of time', seems particularly implacable in Turkana. In many ways the former, the abstract time of sequence and chronology encapsulated by archaeology's 'totalizing narrative' (Lucas 2004: 14), serves to engulf the experiential time of its contemporary population. The particular ways of structuring past, present and future that both inhere in and arise from their involvement with the world around them are perhaps more readily circumvented in the shadow of the abstract chronological calendar manufactured by paleoanthropological and archaeological research. An ostensibly 'traditional' present is more easily assimilated with an unchanging past (a stage in Turkana's measured chronology) than it is explained as a product of its own time and history (cf. Stahl 2001).

This collapsing of recent history into broader narratives of evolutionary time and change is, to a large extent, also a product of a history of visual representation that traces its origins to the earliest colonial conquest of the region, long before palaeoanthropological and archaeological research begun there. Indeed, photography became a critical component in the manufacture of Turkana's timelessness shortly after the turn of the twentieth century, when the first images of Turkana people began to reach Western audiences. Landau (2002: 142) points out that during this time photography had already become one of the most critical 'tools of empire', referencing its role as a ' "scientific" mode of representing human types' (cf. Headrick 1981). Early descriptive photographic catalogues of 'African races' (along with depictions of indigenous Australians and Asians) produced by physical anthropologists entwined seamlessly with a growing scientific interest in biological and social evolution. Such photographs, and the accompanying texts that were distributed throughout the Western world were, for many decades, one of the only means of interaction that members of Western publics had with African populations. Their characteristics – the 'mug-shot'-like poses, orchestrated stances and their arrangement into different categories of dress, ornamentation, ethnicity, race and so on – all worked to reinforce the notions of racial superiority that were fundamental to the colonial project.

In Turkana, despite regular conflict between colonial forces and local communities in the first few decades of the twentieth century, places and people were often photographed by individuals related to the colonial regime (primarily administrative officers). Images from this time are, almost without exception, carefully arranged to include only the most elaborately dressed individuals, and to avoid anything deemed to be 'untraditional'. Such traits can be clearly observed in the photographs of Ernest Emley, one of the earliest and most photographically productive administrative officers to be stationed in Turkana. A photograph taken by Emley in the early 1920s (Figure 10.1) depicts two young women side by side somewhere in Turkana (probably the south). Emley's intention with this image is to portray two distinct styles of dress and body adornment, those of unmarried and married Turkana women. The girl to the left of the photograph, being unmarried, wears a short triangular leather apron, whilst the girl to the right wears skins of a longer length, and extensive ostrich eggshell beads around her waist. Like much of his writing on Turkana, Emley

Figure 10.1 Photograph taken by Ernest Emley in 1921 showing two young Turkana girls. The girl to the left of the image wears the dress and ornamentation of an unmarried woman, whilst the clothing of the girl to the right signifies that she is married (Pitt Rivers Museum 2003.132.3.8).

intended such photographs to simplify and generalize aspects of Turkana culture into categories that might be understood by Western audiences. Their arrangement and accompanying descriptions, if taken at face value, work to disguise variation in such styles across both space and time.

Towards the end of the colonial era and into the decades following Kenya's independence in 1963, photographers who reached Turkana began to opt for more romantic and idealistic representations of the region and its people, a trait that has, to a large extent, endured into the present day and can be seen manifested on the pages of numerous magazines and websites. One of the earliest and most significant bodies of Turkana images to comprise these characteristics is the work of the explorer and travel writer Sir Wilfred Thesiger, who first visited the region in the early 1960s. Thesiger's photographs from this time have been described by Edwards (2010: 106) as 'markers of time and simultaneously a projection into timelessness'. His romantic engagement with the landscape, careful framing and use of light, Edwards argues, work together to endow his numerous subjects with a sense of ancient nobility. Such aesthetic qualities can be recognized in a photograph taken by Thesiger in 1961 (Figure 10.2). The heavily adorned subject, a prominent leader and *emuron* (diviner or seer), is shot from a low angle. He looks down to the lens in a regal pose that is framed only by muted clouds and a soft and low-set horizon.

Figure 10.2 A photograph of an *emuron* (prophet or seer) taken by Wilfred Thesiger on his first visit to Turkana in 1961 (Pitt Rivers Museum 2004.130.30279.1).

Whilst, like Emley's photographs, Thesiger's work is undeniably of great significance and value today, it is important to acknowledge such inventive power and to consider what is left out. His photographs are noticeably far more intimate and relaxed in nature than those of Emley and his contemporaries, and yet they exhibit a similar exclusion of any materials that might be regarded as out of place, including roads, rectangular buildings and 'Western' clothing. In this sense, they seem to concede to the same vision of Turkana that was set out at the earliest colonial contact. In both cases, and indeed in many other photographs of Turkana that have been reproduced for popular consumption more recently, the time of the subjects, as constituted and marked by significant events, processes and changes experienced through the generations, is avoided.

This continues to be the case with numerous representations of the region that arise in the popular press today (McCabe 2004) – the visual tropes that plague East African pastoralist communities (Lane 2015), pitting timeless traditions against the material and technological tide of modernity – seem particularly dogmatic and incontestable in Turkana. What is perhaps most significant is that this continuing evasion of 'Turkana time' seems to correlate remarkably well with the notion of the Turkana Basin as humanity's perpetual birthplace. It is perhaps the case that this conceptualization of the region has only been able to achieve such widespread acceptance outside of northern Kenya because the local population's understandings of time and change have been so persistently and convincingly hidden.

*

The collection that I brought to Turkana in June 2014 comprised over one hundred and fifty photographs. Within it, the early colonial era was represented by Emley's images from the 1920s (e.g. Figure 10.1), followed by a selection of shots dating to the 1930s and 1940s which feature in Pavitt's (2008: 218-19) photographic account of the formation of Kenya colony. Photographs taken by Thesiger on his first visit to Turkana in the early 1960s (e.g. Figure 10.2), portrayed scenes from the middle years of the twentieth century and were followed by images from his last visit in the 1980s. More recent decades were represented by photographs from a range of anthropological and ethno-historical research publications. High-quality versions of all images were printed and brought into the field as a chronologically ordered catalogue, the images that were used were selected to represent a wide range of places, people and objects. For over fourteen months I carried the collection around southern Turkana to a wide variety of villages, towns, marketplaces and homesteads. The photographs were passed around in numerous contexts, from large-scale discussion sessions comprising groups of over fifty people, to more focused one-on-one interviews.

As I noted in the introduction, despite the nature of their fabrication and their inventive vision of Turkana, the photographs in the collection nevertheless inevitably, and usually unintentionally or inadvertently, captured various everyday and mundane aspects of the changing Turkana socio-material universe. It was with the objective of emphasizing these features that I cross-compared the ethnographic photographs with object collections at a range of institutions prior to the fieldwork, and interspersed my own photographs of these objects throughout the larger collection. Using material culture to draw out broader socio-economic, political or environmental changes over long periods of time is, of course, a manner of enquiry that is central to archaeological research undertaken in numerous contexts. Archaeology, as Schiffer (1991: 4) points out, is not only by necessity orientated around everyday things 'but it aims at extracting as much detail as possible about the people who made and used those things found in the ground'. This archaeological predisposition to focus on the detail that inheres in quotidian materialities, however, also aligns with recent trends in the reading of historical photographs in the world of visual anthropology. In particular, Edwards (2001: 2) points out that 'a concentration on content alone, ethnographic appearance – the obvious characteristics of a photograph – is easy, but will reveal only the obvious. Instead one should concentrate on the detail'.

During the photo-elicitation sessions that I undertook in Turkana this focus on material culture, both in terms of the inclusion of my own photographs of ethnographic object collections and the particular questions I pursued, constituted one of the most significant influences that I had on the narratives that emerged. This is not to say that people did not use the photographs for a range of other purposes and to tell a wide variety of other stories that were less concerned with quotidian material histories. To the contrary, many used the collection to name certain deceased individuals, for example, or to trace their own genealogies. The enquiries that I contributed were influenced by all of these other narratives, and by the ways people came to interact with and use the photographs. My interest in changing material culture, although prominent and influential during many exchanges, was thus not passive but rather crafted collaboratively over the months. It was reinterpreted constantly by those who responded to my questions in diverse ways, at times answering with their own questions and at others amending and reorganizing my initial question before replying, but always reforming and remaking the knowledge that slowly accumulated.

Another key influence in the shifting conversations was the form of the collection itself. The affectivity of the photographs did not emerge from their 'volume, opacity, tactility and. . . physical presence' (Batchen, 1997: 2) in a straightforward sense, but rather these qualities seemed to be tightly bound up in the images' representation of many different times and places. In other words, the presentational form of the collection – the chronologically ordered catalogue – was made deeply consequential through the changing styles and colours of visual representation that it comprised. This became clear to me on many occasions, as I observed how individuals made sense of various stylistic transitions, sometimes sorting the images into piles according to the qualities and colours of the photographs and often remarking on the superiority of the more recent images. It was clear that the changing toning and surface effects of the photographs were powerful indicators of the temporal distance between the different albums, and of the passage of time that the overall collection reflected (cf. Edwards 2002).

I would argue that all of these qualities led the collection to evoke accounts of the past that were structured around longer-term narratives of incremental change as opposed to the dichotomy of 'past versus present' or 'then versus now'. Moreover, in making sense of the collection's multi-temporality I did not seek to arrange it around the numbered years of the Gregorian calendar, but rather intertwined it with sequences of key regionally specific past events. Always preceded with the prefix *Ekaru a* (the year of), such events, which range from successful raids of neighbouring groups to periods of heavy rainfall, are commonly used in Turkana to label calendrical years that occurred within living memory. They more rarely also appear in oral histories passed down from previous generations. Sequences of events interweave and imbricate between different livelihoods, villages and regions, and constitute the meshwork upon which both personal biographies and community histories are draped. These biographies and histories are also entangled with and conceptualized through other cyclical notions of time, such as the cultivation cycle of riverside communities (cf. Evans-Pritchard 1939; Shanks and Tilley 1987; Dietler and Herbich 1993; Davies and Moore 2016). This process of relating numerous different timescales is not merely a component of telling stories about the past and narrating history (although it is integral to these activities), but also both emerges through and recursively organizes everyday livelihood practices in Turkana.

Thus, despite their arrangement into chronological order and their depiction of consecutive eras, the photographs did not work to evoke a singular narrative of change over the last century. Instead, those who spent time with the images used them to tell stories that not only related to many different times, but also many different rates and scales of change. Whilst some used them to explore specific memories of objects and locations, or events in which these had been implicated, others opted to narrate much broader changes to society, exchange relations and the environment that have unfolded over several generations. To some, the photographs presented an opportunity to recount their life-story, and yet to others they were a means of outlining the ways in which an entire community had transformed for better or for worse.

*

On a handful of occasions, participants were old enough to recount their own personal experiences of the early years of colonial conquest and indirect rule. This period involved intensive conflict not only between the British and the Turkana, but also various other neighbouring groups. During a large group discussion session held in July 2014 on a market day in Kangarisae (Figure 10.3), Emley's photograph of two young women from 1921, which was discussed above (Figure 10.1), caught the attention of an elderly lady. Elizabeth Epiron had noticed the vast quantity of ostrich eggshell beads that the two women were adorned with. Tracing her finger around the photograph, she explained:

> When I was young, and especially during my mother's life, women wore ostrich eggshell beads around their necks when their husbands were away raiding . . . In those days raiding was [undertaken] with spears not guns. When our husbands returned we would give them the beads to wear around their waists, everyone would know the men had been successful and we would celebrate by slaughtering some of the animals. After some days, the beads would be returned to the women, and we would sew them into our leather skirts. Since guns have come ostriches have gone because hunting them was so easy.
>
> ELIZABETH EPIRON speaking at a group
> discussion session, Kangarisae, July 2014

Figure 10.3 Discussion session with a group of women in Kangarisae in 2015 (photo: Sam Derbyshire).

Epiron's recollection was so vivid and detailed that the discussion that took place over the rest of the session was regularly drawn back to the themes she had initially raised. Conversations concerning livestock raiding in the context of early colonial conquest such as the one that ensued from Epiron's account in Kangarisae, often gravitated toward the point that the British quickly involved themselves in a set of pre-existing dynamics of inter-group warfare in the early twentieth century, staking their power not purely on military imposition or systemic political replacement but rather on livestock's deep entanglement with authority making. It was often recounted that their policy of cattle confiscation meant that they came to be envisioned as raiding partners of enemy groups, such as the Pokot and the Karamojong. A recurring idea was that whilst they ultimately succeeded in their attempts to coerce a large portion of the population into compliance (through the payment of various taxes), they clearly found it necessary to do so largely by incorporating their own authority and objectives into pre-existing conflict and leadership dynamics.

Epiron situated her account in a time when raiding was undertaken 'with spears not guns'. When I returned to this subject in a much later interview with a prominent diviner (emuron) called Lomoru Ima in April 2015 it was pointed out to me that spears were also integral during direct conflicts and skirmishes between the Turkana and the British. Ima recalled that:

> The Turkana were only with spears but the British had guns. Turkana would charge into the British and many would be shot and killed . . . but many would reach the British and the fight would continue at close range with spears.
>
> Interview with LOMORU IMA, Nakurio, April 2015

Aside from the sheer length of time that Turkana communities resisted outright domination using their spears in the face of machine guns, mortars and field guns, perhaps the most compelling detail about their use during this period is that none of them originated in Turkana. During discussions of a selection of Wilfred Thesiger's photographs, and other more recent photographs of Turkana spears from the first half of the twentieth century in Nakaalei and Nakoret, it was explained that both prior to and during the colonial era spears were predominantly acquired via long-distance trade connections with neighbouring groups. In particular, a trade with the Jie and Labwor communities in northern Uganda was outlined on several occasions. Thesiger's images did not date to the time being discussed in these sessions, a point acknowledged by all discussants, but nevertheless served to elicit knowledge of the early colonial era by means of their comparison to it. Other trade routes that had existed alongside and despite the intensive conflict of the early twentieth century were also picked out on various other occasions. One such trade route, which is no longer utilized, was discussed using the example of neck ornaments made from a kind of seed called ngimusio.

These seeds were noted in various different photographs, they were highly valued in the past as their nearest source had been in the upper reaches of the Marakwet Escarpment some 200km south of Turkana.

Through accounts such as these and numerous others it was made clear that whilst widespread violence, cattle confiscations, and escalating warfare on multiple fronts in the early twentieth century were extremely disruptive and distressing in an immediate sense for many of the Turkana individuals and communities who were involved, these activities were also engaged with by means of, and came to be deeply intertwined with, an assortment of prevailing values, practices and systems. These ranged from livestock's deep interfusion with authority and its associated role in political contestations to the centrality and social importance of raiding, and from the long-distance trade networks connecting the Turkana and their numerous neighbours to shorter distance exchanges occurring between communities involved in different livelihoods within Turkana. Whilst most of the political change that occurred during this time was of course unwanted and unwarranted, it is significant that in a pragmatic sense it was not merely impressed upon Turkana society, but rather (at least to a certain extent) emerged through and was crafted by a set of dynamic pre-existing socio-political conditions.

This sense of resilience and continuity underlying punctuated periods of social, economic and political disjuncture did not only emerge during conversations that were focused on early colonial conquest. In discussing the collection in its entirety, several people used it to trace particular social institutions or practices through the years, and to point out that various material and economic changes, such as the shift from using spears during cattle raids to using single-shot rifles and later AK-47s, for example (as mentioned by Epiron), has not necessarily correlated with social or cultural fragmentation. A theme that arose regularly during these considerations was the enduring position of powerful diviners at the heart of Turkana cosmology, and their broader social and political roles. Numerous stories told with the photographs referenced the fact that diviners had been central figures during the initial establishment of *Eturkan* (Turkanaland) long before the years of colonial conquest, and that they had served as key advisors to powerful military leaders during the initial resistance to colonial rule in the early twentieth century. It was also pointed out on several occasions that they had played a vital role in legitimizing the authority of colonially appointed chiefs during the period of indirect rule, and that following independence they continued to be critically influential figures within the politics of intergroup raiding and conflict, regularly using their power to advise the leaders of raiding parties. Often, these accounts of the past roles of diviners were prefaces to descriptions of their enduring political centrality, high social standing and influence, particularly through their entanglement and association with local government representatives.

Figure 10.4 A man from Nakoret holds up an image of his grandfather in 2015, photographed 54 years previously by Thesiger in a nearby location (photo: Sam Derbyshire).

On one occasion, this use of the collection occurred on a distinctly personal level. In a photograph depicting a scene from a meeting in Nakoret (Figure 10.4), a prominent diviner holds a photograph of his own grandfather, known as Lodip, who had been photographed by Thesiger 54 years previously in a nearby location. On discovering the photograph the man explained that, like himself, both his grandfather and father had been highly regarded diviners featuring as key figures in many of the stories regularly recounted amongst the community today. He was most struck by the difference between his own dress, and indeed that of other diviners in the present day, and the ornaments worn by his grandfather. In particular, he took great interest in his grandfather's elaborate head ornamentation which, he explained, he had forgotten despite regularly recounting conversations with and memories of the man. Discussions of changing male head ornamentation, and particularly the headdresses of powerful diviners came to dominate the rest of the session.

Figure 10.5 Emeri Lowasa tells stories about her parents' generation to her eldest son Namoe using Thesiger's photographs in 2015. Her grandchildren look on with interest (photo: Sam Derbyshire).

Whilst large discussion sessions were held regularly throughout the fieldwork, many of the topics that emerged during these sessions were best pursued in smaller, more focused interviews. These interviews were often one-on-one, and conducted with individuals with known knowledge or experience pertaining to a certain topic of interest. Having said this, they were also often undertaken without explicit objectives or research questions and, especially during the latter half of the fieldwork, it became common for people to come to see the collection of their own volition with the intention of exploring the images for their own satisfaction (Figure 10.5). During an interview at a small homestead near to the Lothagam Hills in south east Turkana (Figures 10.6 and 10.7), Emeri Lowasa took particular interest in the difference between Thesiger's photographs from the 1960s and the present day. Many of the women depicted by Thesiger wore large single mollusc shells in prominent positions around their necks. Lowasa pointed out that in many of the more recent images (Thesiger's later shots, and those from more recent research publications), and indeed in the present day, there were very few women with such shells. Instead, women would often wear the modified ends of plastic spoons in a similar fashion as a substitute for the shape and colour of shells. Showing the image to her son, Namoe Lowasa, she explained as follows:

Figure 10.6 Discussion session with a group of men on market day in Nakurio, 2015 (photo: Sam Derbyshire).

Figure 10.7 A woman who walked several kilometres to see the collection after hearing stories at her home, sits and examines photographs from the mid twentieth century, 2015 (photo: Sam Derbyshire).

[There used to be] far less insecurity to the south east of Turkana in the region between here and Isiolo Town. Very few people had guns, and families would frequently travel down to those parts without worry. It was during those trips that people would pick up shells from the dry riverbeds down there, many would bring them back for their relatives, or even trade them. Since my youth that area has become very insecure, there is much fighting between Turkana, Pokot and Samburu. People rarely travel there and therefore we rarely have shells.

Interview with EMERI LOWASA, Nakurio, October 2014

Lowasa's use of the collection to describe the impact of firearms on long-distance movement and trade, connects with Epiron's previously mentioned account of their implication in the disappearance of ostriches and ostrich eggshell beads. In both instances, however, these stories of rupture also came to be embedded in longer-term accounts of continuity and dynamic adaptation. Lowasa outlined recent critical changes in the dynamics of regional conflict that have underpinned a breakdown in the kind of long-distance ambulatory trade that had existed throughout most of the twentieth century, making many of the items that this trade had provided (such as *ngimusio* seeds and shells) inaccessible for the first time. However, her description of the use of plastic spoon heads to replace ornamental shells in female neck ornamentation is also an indication that Turkana communities' outward-looking nature and their dynamic involvement with materials from external regions have not diminished at all. Rather, these aspects have been re-aligned to find a basis within the contemporary circumstances. As various urban centres, such as Lodwar and Lokichar, have expanded over the last few decades the assemblage of new materials and objects that now make their way into Turkana almost solely by means of the Lodwar-Kitale road is, in a wide variety of cases, used to refabricate, reform and restate a diverse array of deeply rooted practices and styles of ornamentation and clothing.

This idea that the gradual growth of urban centres has been utilized by the population of Turkana to negotiate broader economic and environmental changes over the last century also emerged during a discussion session in Nakurio in February 2015. This session (Figure 10.6) mainly comprised male elders from the Nakurio community. It quickly came to be focused on the links between environmental degradation, faunal extinction and male clothing. Marko Eken, a prominent leader in Nakurio, used the collection to deliver the following account:

During the time of our grandfathers, the men in these pictures [points to images from Pavitt's collection of the 1930s], men wore leopard skins and baboon skins over their shoulders. If you were in the *Ngimor* age-set you would wear a baboon skin, if you were a *Ngirisae* you would wear a leopard skin. These men, our fathers [points to Thesiger's photographs from 1961] wore cloth bought from Lodwar, already by this time most wild animals had gone. My father told me that during his father's time the environment was very different, men could hunt and collect wild fruits, and they did not have to travel far away from their families in search of pasture. The world was very different.

MARKO EKEN speaking at a group discussion, Nakurio, February 2015

Eken's story of the shift from animal skins to thick cotton blankets in the 1950s and 1960s led other participants in the group to make the further clarification that in more recent decades cheaper and lighter pieces of cloth, which are made from a mix of cotton and polyester and flow into Turkana by means of its urban centres, have become the most popular materials worn by men. Indeed, most participants in this discussion were wearing such garments (Figure 10.6).

Thus, much like Lowasa's description of the use of plastic spoon heads to replace older forms of female neck ornamentation, the participants in the Nakurio session used the theme of male clothing to outline some of the ways in which various communities have negotiated the growth and expansion of Turkana's urban centres to reconfigure their engagements with external markets, territories and populations. However, the growth of Turkana's urban centres is only one in a long series of broad-scale transformations in which Turkana clothing has been implicated over the years including, as Eken pointed out, the extinction of various wild fauna during the first half of the twentieth century and general environmental degradation in the ensuing years. What is perhaps most significant is that all of these successive processes of ostensibly disruptive transformation were not merely understood as discontinuities in Turkana clothing styles, but rather processes by means of which clothing has been both materially and symbolically reformulated through the years. In other words, the history of clothing, as described during the Nakurio session, does not comprise erosion, discontinuity or disintegration but is rather an articulation of the resilient acts of remaking and reconstituting that underpin and maintain Turkana's complex socio-material networks.

*

The extracts I have discussed above constitute only a small sample of the numerous different themes, historical accounts, genealogies and biographies that emerged during discussions of the collection on its return to Turkana. On the one hand, they clearly reflect the disparity and wide-ranging nature of the accounts that the collection was used to articulate by those who encountered it, the knowledge that emerged over the months of fieldwork was profoundly heterogeneous and, in some cases, opaque. However, this complexity also, to a certain extent, seemed to express an underlying common convergence in conceptions of how social history has been entangled with various transformative episodes over the recent past. What I mean by this is that processes that might otherwise instinctively be correlated with social rupture and discontinuity such as war, drastic economic transformation or faunal extinction, for example, were regularly incorporated into discussions of, or anecdotes pertaining to, longer-term strands of socio-cultural continuity.

This is not to suggest that the accounts that emerged from the collection were all founded on a sense of stasis or perpetuity in Turkana customs, practices or social institutions, but rather that historical change was rarely envisaged as an uncontrolled or unmediated phenomenon derived primarily from outside influences. The prevalent image of Turkana as the realm of the marginalized, the subaltern and the fragile, as might easily be read in the photographs, was thus actively reconfigured and reconditioned through accounts that emphasized productivity, dynamism, ingenuity and, most significantly, social change as a process primarily driven by local individuals and communities. In this sense, many of the stories and narratives that were told using the collection ran counter to what its various albums were initially constructed to portray, and indeed to the enduring colonial legacies of both photography and archaeology that continue to craft interpretations and depictions of Turkana through various media in the wider world. They serve to confute persistent assumptions of passivity and the erosion of tradition, and to emphasize the simplicity and inaccuracy of the notions of 'pessimism and loss' that Henrietta Moore (2011: 5) argues are chief amongst the various presuppositions currently adhering to the idea of globalization.

The idea that an archaeological approach to ethnographic photographs might provide historical narratives that actively subvert and contradict the very vision that both archaeology and photography have co-functioned to reinforce over the years is, on one level, perhaps politically and symbolically significant. It at least has the potential to be so in the future as the project to decolonize archaeological theory and practice continues (cf. Schmidt and Pikirayi 2016), and various museum collections become ever more accessible to source communities across the globe. As both of these processes continue to gather pace, the point that historical photographs tend to become 'symbolic of people's desires to control their own histories and their own destinies' (Edwards 2003: 84) might perhaps instil deeper methodological and theoretical connections between postcolonial African archaeology and visual anthropology. Indeed, whilst the approach I have outlined in this article constitutes just one possible example, and is largely experimental and exploratory in nature, I would suggest that it makes a strong case for the advantages that similar collections of historical ethnographic photographs might bring to archaeological research seeking to engage with African histories.

I do not wish to dedicate extensive space to the question of whether the research I have discussed in this article should be labelled as a form of indigenous archaeology. As I noted above, the connotations of this term's use in African contexts are not the same as those attached to its use in Australia and North America (Lane 2011), and a lengthy debate over its relative merits and faults is perhaps beside the point. I would rather suggest that the most significant questions that emerge from my use of photographs in Turkana relate to the methodological and theoretical implications of close community collaboration at the level of interpretation, a principle that is arguably not limited to indigenous archaeology alone but rather central to the project of postcolonial archaeology in general (cf. Gosden 2012). As I outlined above, I am under no illusions as to the weight of my own influence over the production of knowledge throughout the research I have discussed in this article, nor is my own perception entirely contra to the possibility that archaeology might never fully escape the bounds of Western epistemologies in general, as a result of its own disciplinary heritage. Perhaps it is true that 'the master's tools will never dismantle the master's house' (Lorde 1984: 113). Yet, as Colwell-Chanthaphonh (2012) argues, dismantling the house altogether is perhaps not essential to the task of decolonizing archaeology. We might instead attempt to 'reimagine and remake contemporary archaeology . . . constructing more rooms for more people, new rooms for new views' (Colwell-Chanthaphonh 2012: 282). Such an act of reimagining and remaking will surely only be possible in the house of theory if it also takes place within the very toolset at an archaeologist's disposal, and it is to this latter task that the possibility of historical photographs as archaeological tools is most pertinent.

References

Barringer, T. and T. Flyn, 1998. *Colonialism and the Object*. London: Routledge.
Batchen, G. 1997. *Photography's Objects*. Albuquerque: University of New Mexico Art Museum.
Bell, J. 2003. Looking to See: Reflections on Visual Repatriation in the Purari Delta, Gulf Province, Papua New Guinea. In L. Peers and A. Brown (eds) *Museums and Source Communities*. London: Routledge.
Bell, J. 2008. Promiscuous Things: Perspectives on Cultural Property Through Photographs in the Purari Delta of Papua New Guinea. *International Journal of Cultural Property* 15(2): 123–139.
Bell, J. 2010. Out of the Mouths of Crocodiles: Eliciting Histories in Photographs and String-figures. *History and Anthropology* 21(4): 351–373.
Binney, J. and G. Chaplin 2003. Taking the Photographs Home: The Recovery of a Maori History. In L. Peers and A. Brown (eds) *Museums and Source Communities*. London: Routledge.
Brown, M. 1989. *Where Giants Trod: The Saga of Kenya's Desert Lake*. Long Beach, CA: Safari Press.
Carrier, N. and K. Quaintance 2012. Frontier Photographs: Northern Kenya and the Baxter Collection. In R. Vokes (ed.) *Photography in Africa, Ethnographic Perspectives*. Woodbridge: James Currey, pp. 81–103.

Chirikure, S., G. Pwiti, C. Damm, C. Folorunso, D. Hughes, C. Phillips, P. Taruvinga, S. Chirikure and G. Pwiti 2008. Community Involvement in Archaeology and Cultural Heritage Management: An Assessment from Case Studies in Southern Africa and Elsewhere. *Current Anthropology* 49(3): 467–485.

Clifford, J. 1991. Four Northwest Coast Museums: Travel Reflections. In I. Karp and S. Lavine (eds) *The Poetics and Politics of Museum Display*. Washington, DC: Smithsonian Institution Press, pp. 212–254.

Clifford, J. 1999. Museums as Contact Zones. In J. Clifford (ed.) *Routes: Travel and Translation in the Late Twentieth Century*. New York: Routledge, pp. 188–195.

Collier, J. and M. Collier 1986. *Visual Anthropology: Photography as a Research Method*. Albuquerque: University of New Mexico Press.

Colwell-Chanthaphonh, C. 2012. Archaeology and Indigenous Collaboration. In I. Hodder (ed.) *Archaeological Theory Today*. Cambridge: Polity, pp. 267–291.

Colwell-Chanthaphonh, C., T. Ferguson, D. Lippert, R. McGuire, G. Nicholas, J. Watkins and L. Zimmerman 2010. The Premise and Promise of Indigenous Archaeology. *American Antiquity*, 75(2): 228–238.

Coombes, A. 1994. *Reinventing Africa: Museums, Material Culture and Popular Imagination in Late Victorian and Edwardian England*. New Haven, CT: Yale University Press.

Crooke, E. 2008. *Museums and the Community: Ideas, Issues and Challenges*. London: Routledge.

Davies, M., T. Kipruto and H. Moore 2014. Revisiting the Irrigated Agricultural Landscape of the Marakwet, Kenya: Tracing Local Technology and Knowledge over the Recent Past. *Azania: Archaeological Research in Africa*, 49(4): 486–523.

Davies, M. and H. Moore 2016. Landscape, Time and Cultural Resilience: A Brief History of Agriculture in Pokot and Marakwet, Kenya. *Journal of Eastern African Studies* 10(1): 67–87.

Deetz, J. 1977. *In Small Things Forgotten: An Archaeology of Everyday Life in Early America*. New York: Anchor/ Doubleday.

Denbow, J., M. Mosothwane and N. Ndobochani 2009. Everybody Here Is All Mixed Up: Postcolonial Encounters with the Past at Bosutswe, Botswana. In P. Schmidt (ed.) *Postcolonial Archaeologies in Africa*. Santa Fe, NM: School of American Research Press, pp. 211–230.

Derbyshire, S. 2018. Trade, Development and Destitution: A Material Culture History of Fishing on the Western Shore of Lake Turkana, Northern Kenya. *African Studies* 78(1).

Derbyshire, S. and L. Lowasa forthcoming. The Ruins of Turkana: An Archaeology of Failed Development in Turkana, Northern Kenya. In N. Berre, N. Hoyum, P. Geissler and J. Lagae (eds) *Forms of Freedom: Legacies of African Modernism*. Bristol: Intellect.

Dietler, M. and I. Herbich 1993. Living on Luo Time: Reckoning Sequence, Duration, History and Biography in a Rural African Society. *World Archaeology* 25(2): 248–260.

Edwards, E. 2001. *Raw Histories: Photographs, Anthropology and Museums*. Oxford: Berg.

Edwards, E. 2002. Material Beings: Objecthood and Ethnographic Photographs. *Visual Studies* 17(1): 67–75.

Edwards, E. 2003. Talking Visual Histories: Introduction. In L. Peers and A. Brown (eds) *Museums and Source Communities*. London: Routledge, pp. 83–99.

Edwards, E. 2010. Imagined Time: Thesiger, Photography and the Past. In C. Morton and P. Grover (eds) *Wilfred Thesiger in Africa*. London: Harper Press, pp. 106–115.

Evans-Pritchard, E. 1939. Nuer Time-reckoning. *Africa*, 12(2): 189–216.

Giblin, J. 2010. Reconstructing the Past in Post-genocide Rwanda: An Archaeological Contribution. *Azania: Archaeological Research in Africa* 45(3): 341.

Giblin, J. 2012. Decolonial Challenges and Post-genocide Archaeological Politics in Rwanda. *Public Archaeology* 11(3): 123–143.

Giblin, J. 2014. Toward a Politicised Interpretation Ethic in African Archaeology. *Azania: Archaeological Research in Africa* 49(2): 148–165.

Gosden, C. 1994. *Social Being and Time*. Oxford: Blackwell.

Gosden, C. 2012. Post-colonial Archaeology. In I. Hodder, ed., *Archaeological Theory Today*. Cambridge: Polity Press.

González-Ruibal, A. 2006. The Past is Tomorrow: Towards an Archaeology of the Vanishing Present. *Norwegian Archaeological Review* 39(2): 110–125.

González-Ruibal, A. 2008. Time to Destroy: An Archaeology of Supermodernity. *Current Anthropology* 49(2): 247–279.

Greer, S., R. Harrison and S. Mcintyre-Tamwoy 2002. Community-based Archaeology in Australia. *World Archaeology* 34(2): 265–278.

Harlan, T. 1995. Creating a Visual History: A Question of Ownership. In P. Roalf (ed.) *Strong Hearts: Native American Visions and Voices*. New York: Aperture.

Harmand, S., J. Lewis, S. Feibel, C. Lepre, S. Prat, A. Lenoble and N. Taylor, 2015. 3.3-Million-year-old Stone Tools from Lomekwi 3, West Turkana, Kenya. *Nature* 521 (7552): 310–315.

Harrison, R. and C. Williamson 2004. *After Captain Cook: The Archaeology of the Recent Indigenous Past in Australia*. Walnut Creek, CA: AltaMira Press.

Headrick, D. 1981. *The Tools of Empire: Technology and European Imperialism in the Nineteenth Century*. Oxford: Oxford University Press.

Landau, P. 2002. Introduction: An Amazing Distance: Pictures and People in Africa. In P. Landau and D. Kaspin (eds) *Images and Empires: Visuality in Colonial and Postcolonial Africa*. Berkeley: University of California Press, pp. 1–40.

Lane, P. 2004. Re-constructing Tswana Townscapes: Toward a Critical Historical Archaeology. In A. Reid and P. Lane (eds) *African Historical Archaeologies*. New York: Springer, pp. 269–299.

Lane, P. 2011. Possibilities for a Postcolonial Archaeology in Sub-Saharan Africa: Indigenous and Useable Pasts. *World Archaeology* 43(1) 7–25.

Lane, P. 2015. Sustainability: Primordial Conservationists, Environmental Sustainability and the Rhetoric of Pastoralist Cultural Heritage in East Africa. In T. Rico and K. Lafrenz Samuels (eds) *Heritage Keywords: Rhetoric and Redescription in Cultural Heritage*. Boulder: University Press of Colorado, pp. 259–283.

Leakey, R. and R. Lewin 1979. *People of the Lake: Mankind and Its Beginnings*. New York: Anchor Press.

Lorde, A. 1984. *Sister Outsider: Essays and Speeches*. Berkeley CA: The Crossing Press.

Lucas, G. 2004. *The Archaeology of Time*. London: Routledge.

MacDougall, D. 1997. The Visual in Anthropology. In M. Banks and H. Morphy (eds) *Rethinking Visual Anthropology*. London: New Haven Press, pp. 256–295.

Marstine, J. (ed) 2006. *New Museum Theory and Practice: An Introduction*. Oxford: Blackwell.

McCabe, T. 2004. *Cattle Bring us to Our Enemies: Turkana Ecology, Politics and Raiding in a Disequilibrium System*. Ann Arbor: University of Michigan Press.

Moore, H. 2011. *Still Life: Hopes, Desires and Satisfactions*. Cambridge: Polity Press.

Nicholas, G. 2010. Seeking the End to Indigenous Archaeology. In C. Phillips (ed.) *Bridging the Divide: Indigenous Communities and Archaeology into the 21st Century*. Walnut Creek, CA: Left Coast Press, pp. 233–252.

Nicholas, G. and T. Andrews (eds) 1997. *At a Crossroads: Archaeology and First Peoples in Canada*. Burnaby, BC: Simon Fraser University Archaeology Press (Archaeology Press Publication 24).

Niessen, S. 1991. More to It Than Meets The Eye: Photo-Elicitation amongst the Batak of Sumatra. *Visual Anthropology* 4: 415–430.

Pavitt, N. 2008. *Kenya: A Country in the Making, 1880 – 1940*. New York: W. W. Norton and Company.

Peers, L. and A. Brown (eds) 2003. *Museums and Source Communities*. London: Routledge.

Pikirayi, I. 2004. Less Implicit Historical Archaeologies: Oral Traditions and Later Karanga Settlement in South-central Zimbabwe. In A. Reid and P. Lane (eds) *African Historical Archaeologies*. New York: Springer, pp. 243–267.

Poignant, R. 1994. About Friendship: About Trade: About Photographs. *Voices: The Quarterly Journal of the National Library of Australia* 4(4): 55.

Reid, A. and P. Lane (eds) 2004. *African Historical Archaeologies*. New York: Springer.

Rika-Heke, M. 2010. Archaeology and Indigeneity in Aotearoa/ New Zealand: Why Do Maori Not Engage with Archaeology? In C. Phillips (ed.) *Bridging the Divide: Indigenous Communities and Archaeology into the 21st Century*. Walnut Creek, CA: West Coast Press, pp. 197–212.

Schiffer, M.B. 1991. *The Portable Radio in American Life*. Tucson, AZ: University of Arizona Press.

Schmidt, P. 2010. Social Memory and Trauma in Northwestern Tanzania: Organic, Spontaneous Community Collaboration. *Journal of Social Archaeology* 10(2): 255–279.

Schmidt, P. 2014. Rediscovering Community Archaeology in Africa and Reframing Its Practice. *Journal of Community Archaeology and Heritage* 1(1): 37–55.

Schmidt, P. and I. Pikirayi (eds) 2016. *Community Archaeology and Heritage in Africa: Decolonizing Practice*. London: Routledge.

Schmidt, P. and J. Walz. 2007. Silences and Mentions in History Making. *Historical Archaeology* 41(4): 129–146.

Shanks, M. and C. Tilley 1987. *Social Theory and Archaeology*. Cambridge: Polity Press.

Silliman, S. (ed.) 2008. *Collaborating at the Trowel's Edge: Teaching and Learning in Indigenous Archaeology (Volume 2)*. Tucson: University of Arizona Press.

Spivak, G. 1988. *In Other Worlds: Essays in Cultural Politics*. London: Routledge.

Stahl, A. 2001. *Making History in Banda: Anthropological Visions of Africa's Past*. Cambridge: Cambridge University Press.

Straight, B., P. Lane, C. Hilton and M. Letua 2016. Dust People: Samburu Perspectives on Disaster, Identity and Landscape. *Journal of Eastern African Studies* 10(1): 168–188.

Thomas, N. 1997. Partial Texts: Representation, Colonialism and Agency in Pacific History. In N. Thomas (ed.) *In Oceania: Visions, Artifacts, Histories*. Durham, NC: Duke University Press, pp. 139–158.

Watkins, J. 2000. *Indigenous Archaeology: American Indian Values and Scientific Practice*. Walnut Creek, CA: Altamira Press.

Watkins, J. 2005. Through Wary Eyes: Indigenous Perspectives on Archaeology. *Annual Review of Anthropology* 34: 429–449.

Wood, B. and M. Leakey 2011. The Omo-Turkana Basin Fossil Hominins and Their Contribution to our Understanding of Human Evolution in Africa. *Evolutionary Anthropology: Issues, News and Reviews* 20(6): 264–292.

Yeh, E. 2007. Tibetan Indigeneity: Translations, Resemblances, and Uptake. In M. de la Cadena and O. Starn (eds) *Indigenous Experience Today*. Oxford: Berg.

11
THE AERIAL IMAGINATION

Oscar Aldred

Mid August 2016 – Glenrothes, Fife and above East Scotland. I arrive at Glenrothes airfield with my colleague. We meet up with our pilot, and help take the Cessna 172 out of the hangar. The plane's number GBMCI – or Charlie India, for short – looms on both sides. We prepare the cameras, turn on the GPS to trace our flight, open up the maps, get into the aircraft, belt up and put the headphones on. We listen to the pilot talking to ground control in clipped and punctilious language. In a minute or two, we are airborne, and in a few more we have made our ascent to 2000 ft – our typical operating height.

With each mark on the altimeter and the buffeting from the wind and the turbulence, a new mode of looking is established. At 500 ft the ground moves quickly past, at 1000 ft we begin to see the expanse of a few fields, and at 2000 ft our totalizing gaze of distant horizons and collections of farms and fields comes into view; here we are not too far above to notice detail, but enough distance to identify the next target to photograph. The aircraft environment is a meshwork of different operations and sensations: speed, distance as well as proximity, disorientation during our acrobatic tight orbit flying whilst leaning out of the window with a camera and a map on my lap.

With the ground rushing beneath, the multiple experiences that are a part of this encounter come to my attention; things such as the faint smell of air fuel, the noise of the propeller, and the movements of the plane that touch, collide and enter into the making of the photograph. This is a well-rehearsed hybridity of human and non-human acting as one entity, forging a representational experience like the one I am relating to you here, and the photographs that come to stand in for the experience. However, these experiences are more than representational – they are embodied into the photographs that are made. Here the photograph absorbs multiple bodies, mine, as well as the pilot's and the navigator's, along with the aircraft and its tilt and rotation, the camera and its focus and angle of incidence, the weather, etc. Our collective body act like Csordas's embodiment paradigm; a vessel or tool for perception (Csordas 1990).

*

This chapter explores the practice of aerial archaeology, and the photographs that are taken from above, at an altitude of between one thousand five hundred and two thousand feet (Figure 11.1). It is at this height, in particular, that photography has given archaeologists an unprecedented understanding of the character of archaeological sites, and landscapes, and consequently about the past. But how aerial photography has been considered thus far by archaeology is rarely challenged. What this chapter aims to examine is how photography fits into the performance of aerial archaeology, assessing what we might learn from those that took *slow*, balloon aerial photographs such as Gaspard Felix Tournachon, Henry Negretti and Lieutenant Phillip Henry Sharpe, and the *quicker* early pioneers in aerial archaeology such as Antoine Poidebard, O.G.S. Crawford, G.W.G. Allen and J.K. St Joseph. By adding commentary on the contemporary practice of photography in aerial archaeology, the chapter also looks at the development of its photographic approach, and the way that today aerial photographs are used to convey a different *sense* of landscape from that in the recent past – a sense of landscape in which the temporal component of landscape is less *compressed* into a single image, and considered more expansive.

I want to suggest that there is much more in the aerial photograph than has so far been utilized. The differences are concerned with the obvious extension in

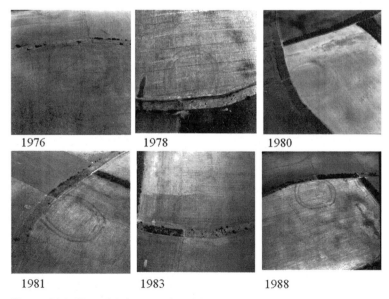

Figure 11.1 Six aerial photographs of the prehistoric fort at Birkenside Hill in the Scottish Borders each showing the fortification in a different light, taken between 1976 and 1988 (Historic Environment Scotland; from top left to bottom right: SC 1550105 (1976); SC 1550106 (1978); SC 1550107 (1980); SC 1550108 (1981); SC 1550109 (1983); SC 1550104 (1988)).

the types of sites that are looked for, guided by an expanded notion of heritage, which for the early, inter-war pioneering aerial archaeologists was narrowly focused on investigating later prehistoric and Roman period monuments. But also how archaeologists have utilized the view from above to convey certain kinds of authority; temporal, spatial or otherwise.

O.G.S. Crawford, the founder and first editor of the world archaeology journal *Antiquity,* writing in 1953, by which time he had practised aerial archaeology for at least thirty years (Crawford 1924), remarked:

> There was not, as some were almost inclined to think, any magic power in the camera; as a matter of fact it sees no more than does the naked eye. What it does is to make a document, the photographic print, which has all the properties of an original historical manuscript except its uniqueness – for it can be replaced if lost. This document can be studied at leisure in the home or office, and compared with others and with maps of the region – things that are impossible when observing from a fast-moving aeroplane.

> CRAWFORD 1953: 46

For Crawford, aerial photographs were like documents, with a materiality and physicality that allowed the archaeological sites and the landscapes in them to be examined 'at leisure'. And they were thus also like maps. Crawford considered an archaeological site when seen from above from an aircraft to be less coherent than when it is seen in a photograph that could be more appreciated and scrutinized for its detail, for an extended period of time. Today the camera technology is digital, and the ways in which aerial archaeologists visualize landscapes have changed too (cf. Brophy and Cowley 2005). Aerial photographs come in a variety of forms. They may be oblique (say, viewing the surface of a landscape at a 45 degree angle) or vertical (viewing the surface of a landscape at a ninety degree angle). These same photographs may have been scanned from photographic film, or may have been born digitally. Satellite images, whether in colour, pseudo-colour or grey panchromatic, have a similar verticality to photographs. Airborne Laser Scan or Light Detection And Ranging (LIDAR) images show a grey-scale digital terrain model or a visualization of a digital land surface. Archaeologists today can digitally navigate these different kinds of images to see the earth from above without moving from the desk, zooming out to trace the outline of the planet and zooming in to the scale of a human in a field. New visual capture technologies are operated by alternative human participation such as remotely operated drones and unmanned aerial vehicles. Digital cameras are GPS enabled, blending viewing with surveying in the landscape. The variety of devices extends new 'ways of seeing' beyond those imagined by John Berger (1977). Aerial images of cities, landscapes, and the planet itself proliferate in smartphone route-planners, advertising, gaming, movies and television news – modes of vision made possible by flight that nevertheless stand in a longer tradition in Western visual culture of views, prospects and bird's-eye views that pre-dates modern aviation technology (Dorrian and Poussin 2013: 1). 'Everyone has a Google Earth story to tell,' suggests Mark Dorrian (2013: 290). There is differential access to these views that maps politically across variances in image resolution through internet speed (Dorian 2013). This super-abundant quality and ubiquitous nature of the aerial image as transformed in the digital world has blurred the boundary between the photograph and the map in a manner that Crawford would have recognized. With access to these new visual modalities, and the attendant changes in perception, we are all aerial archaeologists now.

*

Archaeology is a form of knowledge in which the act of making things visible is central (Hicks Chapter 2 this volume). How do these new modes of visuality affect our knowledge of the past? Consider a tranche of traditional aerial photographs of a single site – the fort at Birkenside Hill in the Scottish Borders, for example, where traces of prehistoric earthworks are visible as a cropmark in a field (Figure 11.1). The monument was first understood from ground surveys undertaken by the Royal Commission on the Ancient and Historical Monuments of Scotland (RCAHMS) for the inventories of 1909 and its revision in 1915. But subsequent narratives about the fort used aerial photographs to add information and change the interpretation of the site. From aerial photographs new features were successively added to the site record description – two concentric ditches, an entrance, two inner-palisaded trenches that suggested an earlier phase in the fort's construction – and new monuments around the fort – a square barrow, a pit alignment, and a rectilinear enclosure.

There is a surprising degree of uncertainty and chance in this visuality, as these features and sites are seen in a photograph in one year, but may be invisible as cropmarks in the next. Archaeologists photographed the site from the air seventy-nine times between 1976 to 2009 (Figure 11.1), and each photograph was captured under different conditions – in terms of the time of day and lighting conditions, the time of year and the type of vegetation or absence thereof, the film used, shutter speed, camera type, framing, angle, altitude and speed of the plane, pilot, operator, and so on. Each element operates to accentuate or not to reveal particular aspects of the archaeological detail of the double-ditched oval fort. These vagaries of photographing of a site from the air bring a partial and provisional quality to each scientific representation, as knowledge is made through a transformation of the landscape into a partial image, rather than documented through an all-seeing vision. Acknowledging the 'ecology' of practices involved in producing aerial photographs, from farming to flying, is to acknowledge that there is no magic power in images of the past made with an airborne camera.

This observation has implications for the everyday aerialities of modern vision, which must themselves like the archaeologist's photographs always be images of the near past rather than the immediate present (even if there is only a small delay in the case of live images), acting as deferrals of some kind. Our cultural saturation of aerial imagery has de-sensitized us to the partiality of the view from above, leading us to forget that it is always a view from afar.

Aerial photographs are normally used to tell quite conventional stories about archaeological sites and landscape. But the aerial view continues to enter an archaeological consciousness in a number of other ways (Figure 11.2). Instead of a cartographic map being used to show a site location in an archaeological report on an excavation, an aerial photograph or satellite image is often used. This mode of seeing, some might say, is an extreme instance of what Donna Haraway called the 'god trick' of appearing to adopt a universal gaze (Haraway 1991; Hicks and McAtackney 2007), but as it has entered the archaeological consciousness it is beginning to replace long-established conventions, altering how we envision the site and its unbounded landscape context. This is not a problem in itself – and actually quite useful to break down the arbitrary divisions between a site and its landscape – but what this is leading to is an uncritical use of the view of above, ignoring assumptions that underlie what such a view from above does. These kinds of images therefore come to stand in for the object of our interest. As with cartography, the landscape becomes an abstraction, a

Figure 11.2 The sortie as an experience that is shared between colleagues as a hybrid system (photo: Oscar Aldred).

stylized image that extends reality (cf. Shanks 2012). The stylized and privileged position conveys a specific view; it communicates the author's intention (e.g. Crawford and Keiller 1928; Griffith 1988). With the proliferation of views from above, the status of archaeological reality has changed. If we are so used to seeing an archaeological site from above, we forget that 'the map is not the territory' (Korzybski 1933: 58), so that there is no 'difference between the world as we see it and the world as it actually is, beyond our faulty memories and tired understanding' (Rausing 2015). But can aerial photography reclaim something of the older, slower mode of reflection that interested Crawford, by aligning it more fully with an archaeology that combines theory and practice (cf. Lucas 2012)?

There has been some discussion on this matter already. Kitty Hauser's *Shadow Sites* (2007) explored Crawford's attitudes to photography and landscape with reference to Modernists and Neo-Romantic artists such as John Piper, Paul Nash and Geoffrey Grigson. Hauser's account focuses on the familiar idea of the photograph as a moment of time captured as 'proof of existence of the sites, artefacts, and practices' (Hauser 2007: 118). She traces how the Neo-Romantic tendencies were inspired, enhanced and communicated through archaeological aerial photography, alongside the more 'artistic' exchanges represented by Jacquetta Hawkes' *A Land* (1951), suggesting that 'aerial archaeology can be seen not only as an archaeological tool, but also as the embodiment of a kind of historiography that is both materialist and redemptive' (Hauser 2007: 176).

Are we entering a new phase of change in these materialist and redemptive qualities of aerial photographs? As I have suggested, the aerial photograph is an end product in a hybrid system or ecology that is never neutral. Such ecologies are subject to change. But the photograph is also subject to being read, which can have implications for scale. Here distance has a paradoxical function. The closer to the ground, the more detail can be seen, but because there is the appearance of moving much quicker than when at a higher altitude, the view appears less connected to the archaeology. Thus, at a higher altitude and a further distance from the ground, with the illusion of a slower speed, it is possible to feel closer to what can be seen because more coherent detail can be seen. It is possible, too, to gain this sense from interrogating an aerial photograph with a magnifying lens, or with the new technological ability to zoom in to a digital image.

*

Aerial photographs have been used by archaeologists to interrogate site complexes and landscape since the early twentieth century (e.g. Crawford 1923; Crawford and Keiller 1928; Bradford 1957; Wickstead and Barber 2012). But while aerial photographs and aerial images are becoming commonplace and changing through digital technology, the ways that archaeologists think about them have hardly changed. Despite the increased familiarity and utility of aerial photographs, their use in archaeology has often remained, as Matthew Johnson has argued, as artefacts of Romanticism, as the aerial photograph becomes an aestheticized object used to tell an anecdote beyond the site (Johnson 2007: 89–93). Johnson, following arguments familiar from the work of Julian Thomas and Chris Tilley (e.g. Thomas 1993; Tilley 1994), argues that the aerial photograph foregrounds the larger patterns that we can see in space and time, but also establishes a spatial view that privileges elevation where 'the human realities of small-scale movement on a human pattern are lost' (Johnson 2007: 93). The Romantic affinities noted by Hauser in the work of Crawford are certainly an element of how aerial photography emerged as a new technology for viewing landscapes archaeologically (Crawford 1921), but these tendencies are not an inevitable state of affairs when we consider the operational practices of working with aerial photography on a day-to-day basis (Brophy and Cowley 2005; Wickstead and Barber 2012).

The aerial photograph in Figure 11.3 was taken on the 2 December 2004 to evoke the Scottish landscape's character, with low light and incoming cloud from the west, with the hills protruding above. In order to understand the system behind the photograph (after Flannery 1967: 120) and reassert the type of archaeological value contained in a photograph and its theoretical implications, a certain amount of new reflection is needed. The photograph in this embodied,

Figure 11.3 General oblique aerial view looking West across the Gelly Burn, Perth and Kinross, Scotland, towards the Cleish Hills (Historic Environment Scotland DP 008007).

representational and experiential fusion does not freeze the moment of observation in the midst of a tension. It might seek to convey a particular ideology concerning the past. But there is more temporal complexity since even though photographs may be captured in measurable moments of time, they are never conceived in that moment; they emerge through what Patrick Keiller calls 'the unusual, unexpected experience that produced the photograph' (Keiller 2014: 184).

The aerial archaeological photograph is formed through a multiplicity of *accumulated* times that are captured in a moment when pressing the camera button. This accumulation of times come in several forms, from the materiality of the things being photographed to the number of times a site is observed from the air. Many archaeological sites are photographed within a particular season when the cropmarks are clearer, or year on year. And there are many visits where photographs are not taken, but are just observations – which do not remain confined to the unconscious, the 'inner experience of an alienated and rather unreliable artificial reality' (Keiller 2014: 79), but are another layer in the *accumulation* through which the landscape becomes visible in the aerial photograph. The sequence of observation, navigation, dialogue between photographer and pilot, the closing of the camera shutter, the influence of previous observations, even the path between the retina and the brain, form a kind of 'operational chain' (Leroi-Gourhan 1993). Photography becomes part of the time-depth of landscape, as the accumulation of material and embodied actions through time (Figure 11.4) – an observation that relocates the corporeal and sensory Romanticism of seeing landscapes through temporal complexities.

Figure 11.4 Oblique aerial view of the Ring of Brodgar, Orkney, Scotland, 1965 (Historic Environment Scotland SC 345535).

When I think about what my colleague and I have photographed during a flight's sortie, the common assumption of landscape phenomenology that archaeological recording like aerial photographs creates a distance between ourselves and landscape. But the movement of the aircraft and us in it is no less an embodied experience than a journey on foot: it is no less of a visit to the site. As J.K. St Joseph, one of the pioneers in aerial archaeology, wrote in in December 1944, distance increases perception:

> An air photograph shows features viewed comprehensively without the restricted vision of those too near the object; the view is sufficiently distant to remove the distortion induced by perspective. Moreover an air photograph records not only the conventionalized plan of a countryside, as is drawn on a map, but also its actual state, including therefore not only real roads, railways, cottages, and towns, but also the crops, hedgerows, and ephemeral details which can never be distinguished on the most careful, general survey.
>
> ST JOSEPH 1945: 47

However, there are advantages to each viewpoint and mode of encounter with the past. The differences between the aerial photograph and the site visit on foot are not simply due to spatial distance, perspective or from the physicality of touch. When we ask what an aerial photograph of a monument like the Ring of Brodgar (Figure 11.4) does, then we need to think about actions that range from prehistory to the current fieldwork ongoing at the site.

*

'It is the artist who tells the truth and the photograph that lies. For in reality, time does not stand still.' In her critical history on Eadweard Muybridge and his motion capture photography, Rebecca Solnit (2004: 196–8) has suggested that a nineteenth-century conflict can be seen in this comment by Rodin. We perhaps witness the same conflict today between what an aerial photograph captures at high speed and what the naked eye sees, but Rodin's comment can be understood as concerned not only with how a photograph misrepresents, but also with what it is capable of. Muybridge's late-nineteenth century photography advanced technological innovations such as the automated shutter and made it possible to think about the experience of movement in a new way. The mystery of whether a horse flies when it runs was broken, in a moment of change in the comprehension of the everyday that affected what Lefebvre called the 'structure of feeling' (Lefebvre 1991: 251) (Figure 11.5).

The aerial photograph shifted the structure of feeling about archaeological landscapes in an equally profound way that we now take for granted, and further developments in technology continue the process. What changes will be

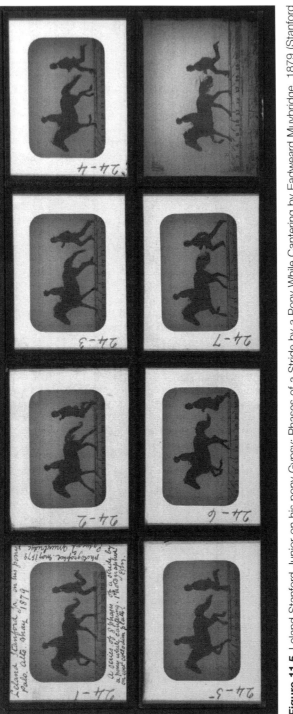

Figure 11.5 Leland Stanford Junior on his pony Gypsy: Phases of a Stride by a Pony While Cantering by Eadweard Muybridge, 1879 (Stanford Family Collections, Cantor Arts Centre, Stanford University).

brought, for example, by the ARGUS camera's Wide Area Persistent Stare imaging system that can be attached to unmanned aerial vehicles and can use hundreds of mobile phones at one time in a mosaic and can have a capture area of up to one hundred square kilometres, with a ground sample pixel size of fifteen centimetres from an altitude of five kilometres (Sloggett 2014)? The aerial imagery that is produced allows forensic and searchable playback in a period of exposure *after* the events that the camera has captured, tracking objects and people. The detail of events can thus be experienced even after the events themselves have unfolded like a video recording that can be watched again and again – and maybe also like an archaeological archive that can be returned to in the museum. We can only imagine the transformation in the structures of feeling, and the politics of privacy and surveillance, in how we perceive the world from above that such 'archaeological' and aerial developments in visual technology will bring.

Filmmaker Patrick Keiller has explored the relationships between technology and visual reality by altering and advancing the spatio-temporal dimensions of photographic and film capture. Keiller (2014: 184) examines the implications of Raoul Vaneigem's assertion, in *The Revolution of Everyday Life,* that

> Not everything affects me with equal force, I am always faced with the same paradox: no sooner do I become aware of the alchemy worked by my imagination upon reality than I see that reality reclaimed and borne away by the uncontrollable river of things.

VANEIGEM 1972 [1967]: 236

For Keiller, a photograph provides a link between the experience of the world and its decay through our imagination of it. It is a transformation because it is a reference to a thing in reality but is itself the product of a new type of reality. While it might seem as if a photograph is a reduced version of the experience of a thing, it equally amplifies certain properties of it. Vaneigem's dilemma is partly resolved through the production of photographs because instead of the 'uncontrollable river of things' impeding on the imaginative pursuit, it holds on to some of the imaginative qualities that may gradually be lost because of the fragility of human memory.

When we ask what aerial photographs do, we raise the question of the human position in what Susan Sontag called the 'image-world'. For Sontag, a photograph appropriates the world. It documents sequences of consumption, furnishes evidence, and can be used for surveillance, while also certifying experience and giving the appearance of participation. Photographs distort the world because they not only fiddle with its dimensionality, but because the act of taking a photograph represents a kind of social rite, a defence against anxiety, and a tool of power (Sontag 1973: 1–8). It is these processes through which 'reality is confirmed and experience enhanced' (Sontag 1973: 24) that framed Sontag's notion of the image-world as a kind of hyper-reality alongside the dulling of the senses:

> to possess the world in the form of images is precisely to re-experience the unreality and remoteness of the real.
>
> SONTAG 1973: 164

So, from Keiller we gain a sense of the photograph facilitating the continuity of reality, while from Sontag we gain the sense of enhancing the experience of reality. Archaeologists, meanwhile, tend to look at a photograph as an unreal deferral of the real, through its document-like properties and their own archival practices. But the archaeological photograph performs more than just the suspension of the real.

We do not need the Benjaminian notion of 'optical unconsciousness' (Benjamin 2008 [1935]: 37), or the cultural anthropologists' account of the pro-filmic, to observe that the archaeological photograph contains more than just the subject of the shot, or the objective of the photographer. Archaeology, after all, is knowledge made through the vagaries of survival – a more-than-momentary perceptive vision that is more macula imprint that lingers on after the flash-image burns itself onto the retina than it is a fleeting illumination. And not just an accumulation: archaeological photography is an exercise in re-imagination.

Consider for a moment the potential of unhinging of the type of archaeology that landscape phenomenology has advocated, focusing on how the sensual experience of the world around constitutes the point of departure for our understanding the past (Tilley 1994, 2010), from a definition of archaeology as the representation of the past. The potential is for the non-representation, including the archival and the photographic, to represent another type of experience, and perhaps even an extension to it, or protention of it (cf. Hicks 2016). Such a consideration brings with it the possibility of looking at an aerial photograph afresh, extending the archaeological sub-field 'landscape archaeology' as a type of post-phenomenology or a philosophy of science (e.g. Ihde 1993; Ihde and Selinger 2003). This consideration is to recognize how the materiality, temporality and imaginative properties of objects can go beyond the representational purposes for which they were made.

O.G.S. Crawford is credited for bringing aerial reconnaissance of archaeological sites to Scotland with a series of pioneering flights in the 1930s that were published in the journal *Antiquity* (Crawford 1930, 1939; Cowley and Gilmour 2005: 51–5) (see Figure 11.6). But of course Crawford's public academic accounts of aerial archaeology rarely include the detail of operational matters that occurred during sorties. Crawford's private archive reveals a different

Figure 11.6 Near vertical aerial view of the Temporary Roman Camp at Grassy Walls, Perth and Kinross, Scotland by O.G.S. Crawford, circa 1939 (Historic Environment Scotland, SC 1039480).

practice. In written correspondences with J.K. St Joseph in 1950s, now held in the Historic Environment Scotland's National Record collections and in the Ashmolean Museum at Oxford University, a great deal of what the experiences of flying were like and what went into taking aerial photographs is made clear – as well as the meaning that was derived from them at the time. Crawford highlights the personal nature of the decisions that were made during an aerial survey flight, influencing what was photographed or not, and what the view from above revealed about the landscape, in particular, Romano-British camps and roads. For example, Crawford apologizes for the 'slight but inevitable intrusion of the personal element imposed by the character of the narrative' when describing a photograph (Crawford 1930: 273; cf. Cowley and Gilmour 2005: 51; Crawford 1939: 289).

The normally hidden but crucial details range from who the photographer was (Did someone else besides Crawford take the photographs? Where did they sit in the plane? What was their intention?), to how the photograph was subsequently used and read. Crawford's letters underline how an aerial photograph is never created in a neutral and objective way, and its version of reality stands in for the things that are in the photograph. But the late twentieth-century post-processual concerns with the construction of the past through its representation were too often based on a naive misrepresentation of the more complex practices of field archaeology as an exercise in 'thick description' (Geertz 1973: 6), rather than, for example, Ruth Benedict's understanding of experience as human imagination (Benedict 1971: 16).

*

Some concluding thoughts. Archaeology is like photography in reverse; it reveals durations of past times; it enacts a 'present that endures' (Bergson 2007 [1946]: 127). We need to pay more attention to how photographs displace and consume time. Like all archaeological photography, the aerial photograph visualizes the past through a different shutter speed and duration of exposure and imprinting. The metaphor of archaeology as a camera is concerned with the discipline's process that works with light. Exposure in photography may evoke ideas of the moment or a chemical process, but in archaeology exposure also exists in revealing past surfaces and features that may have been hidden from the light for thousands of years, whether excavated or revealed from the air as crop marks. In an aerial photograph archaeological features return our gaze as if conveying generations of moments through time. This looking back is reminiscent of Jacques Derrida's sense of *spectre*:

> But what else can you do, since it is not there, this ghost, like any ghost worthy of the name? And even when it is there, that is, when it is there without being there, you feel that the specter is looking, although through a helmet; it is watching, observing, staring at the spectators and the blind seers, but you do not see it seeing, it remains invulnerable beneath its visored armour.
>
> DERRIDA 1994: 124

This sense of an archaeological feature staring back at us is also evoked by Berger's notion of seeing as a kind of reciprocity, whereby if one stands in a landscape, on a hill looking at another, there is always the possibility that if another person is standing on the hill being looked at, she will see you (Berger 1977). Archaeology, as seen from the air or on the ground, is camera-like because it makes it possible for the archaeological feature, site, or landscape to look back at you for an extended period of time, providing another afterlife for something of the embodied histories of past people and their entangled material remains. For the envisioning of new pasts.

References

Benedict, R. 1971. *Patterns of Culture.* London: Routledge and Kegan Paul.

Benjamin, W. 2008 [1935]. *The Work of Art in the Age of Mechanical Reproduction* (trans. J.A. Underwood). London: Penguin.

Berger, J. 1977. *Ways of Seeing.* London: Penguin Books.

Bergson, H. 2007 [1946]. *The Creative Mind: An Introduction to Metaphysics* (trans. M.L. Andison). New York: Dover.

Bradford, J. 1957. *Ancient Landscapes.* London: G. Bell and Sons.

Brophy, K. and D. Cowley (eds) 2005. *From the Air: Understanding Aerial Archaeology.* Stroud: Tempus.

Cowley, D. and S.G. Gilmour 2005. Some Observations on the Nature of Aerial Survey. In K. Brophy and D. Cowley (eds) *From the Air: Understanding Aerial Archaeology.* Stroud: Tempus, pp. 50–63.

Crawford, O.G.S. 1921. *Man and His Past*. Oxford: Oxford University Press.

Crawford, O.G.S. 1924. *Air Survey and Archaeology*. London: HMSO, Ordnance Survey Professional Papers (New Series) 7.

Crawford, O.G.S. 1923. Air Survey and Archaeology, and Discussion. *Geographical Journal* 61(5): 342–366.

Crawford, O.G.S. 1930. Editorial. *Antiquity* 4: 272–278.

Crawford, O.G.S. 1939. Air Reconnaissance of Roman Scotland. *Antiquity* 13(51): 280–292.

Crawford, O.G.S. 1953. *Archaeology in the Field.* London: Phoenix House.

Crawford, O.G.S. and A. Keiller 1928. *Wessex from the Air.* Oxford: Clarendon.

Csordas, T.J. 1990. Embodiment as a Paradigm for Anthropology. *Ethos* 18(1): 5–47.

Derrida, J. 1994. *Spectres of Marx: The State of the Debt, the Work of Mourning, and the New International* (trans. P. Kamuf). London: Routledge.

Dorrian, M. 2013. On Google Earth. In M. Dorrian and F. Pousin (eds) *Seeing from Above: The Aerial View in Visual Culture*. London: Tauris, pp. 290–307.

Dorrian, M. and F. Poussin 2013. Introduction. In M. Dorrian and F. Pousin (eds) *Seeing from Above: The Aerial View in Visual Culture*. London: Tauris, pp. 1–10.

Flannery, K. 1967. Culture History vs Culture Process: A Debate in American Archaeology. *Scientific American* 217: 119–22.

Geertz, C. 1973. Thick Description. Toward an Interpretive Theory of Culture. In C. Geertz *The Interpretation of Cultures. Selected Essays by Clifford Geertz.* New York: Basic Books, pp. 3–30.

Griffith, F. 1988. *Devon's Past: An Aerial View*. Exeter: Devon Books.

Haraway, D. 1991. Situated Knowledges: The Science Question in Feminism and the Privilege of Partial Perspective. In *Simians, Cyborgs, and Women: The Reinvention of Nature*. Free Association Books, London, pp. 183–201.

Hauser, K. 2007. *Shadow Sites: Photography, Archaeology, and the British Landscape, 1927–1955*. Oxford: Oxford University Press

Hawkes, J. 1951. *A Land.* London: Cresset Press.

Hicks, D. 2016. The Temporality of the Landscape Revisited. *Norwegian Archaeological Review* 49(1): 5–22.

Hicks, D. and L. McAtackney 2007. Landscapes as Standpoints. In D. Hicks, L. McAtackney and G. Fairclough (eds) *Envisioning Landscape: Situations and Standpoints in Archaeology and Heritage*. Walnut Creek, CA: Left Coast Press (One World Archaeology), pp. 13–29.

Ihde, D. 1993. *Postphenomenology: Essays in the Postmodern Context.* Evanston, IL: Northwestern University Press.

Ihde, D. and E. Selinger (eds) 2003. *Chasing Technoscience. Matrix for Materiality*. Bloomington and Indianapolis: Indiana University Press.

Johnson, M. 2007. *Ideas of Landscape.* Oxford: Blackwell.

Keiller, P. 2014. *The View from the Train: Cities and Other Landscapes*. London: Verso.

Korzybski, A. 1933. *Science and Sanity: An Introduction to Non-Aristotelian Systems and General Semantics*. Lakeville: International Non-Aristotelian Library.

Lefebvre, H. 1991. *Critique of Everyday Life*. (trans. J. Moor). London: Verso.

Leroi-Gourhan, A. 1993. *Gesture and Speech* (trans. A. Bostock Berger). London: MIT Press.

Lucas, G. 2012. *Understanding the Archaeological Record*. Cambridge: Cambridge University Press.

Rausing, S. (ed.) 2015. The Map Is Not the Territory. *Granta: The Magazine of New Writing* 131.

Shanks, M. 2012. *The Archaeological Imagination.* Walnut Creek, CA: Left Coast Press.

Sloggett, D. 2014. *Drone Warfare: The Development of Unmanned Aerial Conflict*. Barnsley: Pen and Sword.

Solnit, R. 2004. *River of Shadows: Eadweard Muybridge and the Technological Wild West*. London: Penguin.

Sontag, S. 1973. *On Photography.* New York: Farrar, Straus and Giroux.

St Joseph, J.K. 1945. Air Photography and Archaeology. *The Geographical Journal* 105(1): 47–59.

Thomas, J.S. 1993. The Politics of Vision and the Archaeologies of Landscape. In B. Bender (ed.) *Landscape, Politics and Perspectives*. Oxford: Berg, pp. 19–48.

Tilley, C. 1994. *A Phenomenology of Landscape: Places, Paths, and Monuments*. Oxford: Berg.

Tilley, C. 2010. *Interpreting Landscapes: Geologies, Topographies and Identities*. Walnut Creek, CA: Left Coast Press.

Vaneigem, R. 1972 [1967]. *The Revolution of Everyday Life* (trans. D. Nicholson-Smith). London: Practical Paradise.

Wickstead, H. and M. Barber 2012. A Spectacular History of Survey by Flying Machine! *Cambridge Archaeological Journal* 22(1): 71–88.

12
THE TRANSFORMATION OF VISUAL ARCHAEOLOGY (PART TWO)

Dan Hicks

Five keywords accumulated through the argument of the first part of this discussion (Chapter 2 above): *Archaeography, Idiorrhythmy, Histories, Visualism*, and the idea of the knowledge that emerges through Archaeology's visualism – *Photology*. To begin assembling a toolkit for Visual Archaeology, this second part of my account of 'the transformation of visual archaeology' puts forward five further words, seeking not to match up to the 'Seven Theses for Photography' set out by visual anthropologist Chris Pinney (Pinney 2012), but to imagine a new active vocabulary of five inflections – or, perhaps better, five *Mythologies* to join those keywords and to put the notion of photological knowledge into practice: *Homonymy, Prefiguration, the Unhistorical, Impermanence, Appearance*.

VI. Homonymy

Low sunlight and long shadows fall across a poorly-drawn provincial football pitch near King's Lynn, Norfolk in David Wilson's vertical air photograph (Figure 12.1) taken on January 3 1970 and reproduced in his textbook *Air Photography for Archaeologists* to reveal an imperfect reality:

> The incorrect angles, curves and distances incorporated in the layout of this pitch were evidently far from obvious on the ground, yet from the air they are painfully apparent.
>
> WILSON 1982: 15

How to understand the knowledge constituted by this image? Should we seek to relativize this effort at calculated verisimilitude, the conceit of a camera strapped

Figure 12.1 A football pitch south of King's Lynn, Norfolk, 3 January 1970. Vertical photograph (scale 1:1200). From Wilson (1982: 15).

to the fixed wing of a plane to create a measured, vertical distance between the airborne field scientist at one vertiginous extreme and at the other some lines of white paint, traces of skew-whiff hand-wheeling clinging to blades of grass? Historicize the resulting image, calling it a leftover from nineteenth-century empiricist ideals of 'noninterventionist objectivity' (Daston and Galison 1992: 98)? Assert that the image is an object? ('Painting is not photography say the painters. But photography is also not photography, which is to say, it is not a copy', wrote René Crevel; but what then of a photograph of painted turf (Crevel 1925: 23)?)[1] Classify it as just another perspective on the world, an interpretation based on the fake construction of a disembodied gaze, the impossible fiction of the view from nowhere (as if 'the visible world is arranged for the spectator as the universe was once thought to be arranged for God', as Berger wrote (Berger 1972: 16))?

But in adopting such a critical, contemplative reading, in seeking to reveal, so to speak, the subjectivity of the objectivity, in standing back to contextualize, we stand – stand with this secondary, hermeneutic kind of detachment in mind – where exactly? Not nowhere of course, or somewhere on the ground, or hovering in mid-air clinging to the line of sight, but somehow (impossibly) *anywhere* – to view the multiplicity of potential meanings in which the photograph as an enactment of perspective will be repeatedly refracted through human standpoint, and/or cultural perspective, and/or social context, etc., as if each substituted for the photographer's choice of frame, filter, lens, lighting, angle or point of view.

No. Surely it must be possible to imagine a third kind of detachment, neither the lofty empiricism of the air photographer nor the unlocatable relativism of the archaeocritic, which steps back from any false choice between description and embodiment. Is it possible to frame the problem of this photograph as one not of measured substitution or mimetic simulacrum, but of the performance of transformation? Silverman (2015) has the idea that we could understand photographs not so much as indexes, or representations, or copies of reality, or traces, but analogies; but such is the artificial nature of the photological knowledge that emerges through Visual Archaeology that it surely confuses subject and object positions, making the suspended camera part of what is depicted, in a manner that dislocates the referent. It triangulates analogy. Hardly a mimetic method then, but surely not an analogical one either. We would be hard pressed to reconcile even Chris Pinney's ethno-Latourian account of camerawork as a 'technical process' connecting humans and objects as mutual actors in processual networks (Pinney 2010: 148, 150) with this blurring of actants, confusions effected between the technical, cultural and natural worlds. Recall Barthes:

> Originally photographic equipment was made with the techniques of cabinetmaking and precision mechanics: cameras were essentially clocks for seeing, and perhaps within me someone very old still hears in the camera the living sound of the wood.

BARTHES 1980: 33[2]

Photography, through what André Breton described as its '*mechanizing to the extreme the plastic mode of representation*' (Breton 1969 [1935]: 272),[3] has a technical effect upon knowledge that is akin to David Clarke's classic account of how radiocarbon dating transformed the archaeological past (Clarke 1973: 10). 'The archaeologist must now think directly in terms of the kinked and distorted time surfaces of the chronometric scales which he actually uses,' he wrote. Distortions are discovered on the face of the world through the technical application of verticality. Could we imagine this discovery as neither a privileged depiction nor a misrepresentation to be deconstructed, but as a kind of transformation? A precise visual measurement that does not just alter our perspective but skews the marks of the lived environment too. If the camera represents, as Barthes has it, a 'clock for seeing', then photological knowledge is surely concerned not just with the marking of unilineal time, but in the marking of football pitches? Concerned with transformations not just with the measurement of white paint traces but with somehow re-tracing a thread of light back through its diffusion, so that it does not find understanding in terms of something else, but rather produces knowledge in the form not of synonyms but of homonyms. In a manner akin to the 'equivocation' that Eduardo Viveiros de Castro (2004: 5) identifies in indigenous perspectivism: 'the referential alterity between homonymic concepts'. It is as if photological knowledge could recursively distribute Salvador Dali's account of the 'double image' through the act of casting light, of discovery: 'a representation of an object which, without the slightest figurative or anatomical modification, becomes the representation of another totally different object' (Dali 1930: 10).[4] Forty-five minutes each way.

*

VII. Prefiguration

It is, or it was, May 1993 and my 20-year-old body is half way along a trench (Figure 12.2) cut across a ploughed field in south Warwickshire, part of an archaeological evaluation in advance of the construction of the Norton-Lenchwick Bypass. Behind me are, or at least were, Kevin and Sarah, and behind Sarah is Tim who's wearing a hard hat. A steel surveying arrow secures the measuring tape that runs along the centre of the trench from which the ploughsoil has been scooped. The trowel in my hand is tracing the edge of a Romano-British ditch, distinguishing the different shades of layers and fills of different dates of

Figure 12.2 Evaluation trench for the A435 Alcester-Evesham Bypass, Warwickshire, May 1993. The excavator nearest to the camera is the author (Photo: John Thomas).

deposition, in anticipation of the box graders and bulldozers that would construct the dual carriageway (opened August 1995). The exercise in which our group, museum workers in the field, is engaged is one of 'preservation by record'. It involves the destruction of in-situ remains carried under archaeological conditions, sampling and archiving what will be erased.

> Not to know what one is going to discover is self-evidently true of discovery. But, in addition, one also does not know what is going to prove in retrospect to be significant by the very fact that significance is acquired through the subsequent writing, through composing the ethnography as an account after the event. The fieldwork exercise is an anticipatory one, then, being open to what is to come later.
>
> STRATHERN 1999: 9

Some months ago, and perhaps here I must blame the instinctive retrospection and self-regard of my disciplinary instincts, I came across this photograph of myself catalogued alongside the pot sherds and bone fragments in a museum archive; an extreme instance of Proustian *mémoire involontaire*, or Benjaminian optical unconsciousness, or some other hyper-reflexive type of self-haunting, no doubt.[5] No straightforward autobiographical trace simply enduring through the archive in a life that runs parallel to my own, this confusing reunion is more *autopoiesis* than flashback or reconciling denouement, a *mathesis* (institutional ordering) in which we might identify the photographic archive with myths and music as they appear in the account of Lévi-Strauss, and thus with the trowel and the bulldozer alike: as 'machines for the deletion of time' (Lévi-Strauss 1964: 23–34).[6]

If a photological perspective reveals the photograph not as material remnant or trace but as human thought and knowledge, then to discover in the archive one's own anticipatory performance of the human past is surely to encounter some form of prefigurative thought, emerging in what Barthes describes as

> that very subtle moment when, to tell the truth, I am neither a subject nor an object but rather a subject who feels himself becoming an object: I am then living a micro-experience of death (of parenthesis): I really am becoming a ghost.
>
> BARTHES 1980: 30[7]

What has my body become? Not exactly an artefact present in an archive, nor just a concept, nor a multiplication, but some kind of nonmimetic distribution. A transformation. (Barthes: 'Photography every now and then makes what you never perceived in a real face (or reflected in a mirror) appear.' [Barthes 1980: 160–1][8]) The language of absence and presence is inadequate for the description of those blue jeans, the 4AD T-shirt, those DM boots, sunburnt arms, that hand (my past hand) pressing into Roman soil – flattened prosthetics. Surely John Berger cannot be right when he writes (1976: 82) that 'a photograph is static because it has stopped time'? This image is much less a trace or a remnant than it is some medium of framed regression. Resemblances stretch across time with that Duboisian 'elasticity' (Dubois 2016: 164), strung out from the Barthesian punctum (coming face to face) back to Muybridge's chronophotography, but constantly springing back in some illuminated reversal of Benjaminian history. A storm of returning that neither awakens the dead or deadens the living. Inalienable salvage. The flash of the archaeologist's/this archaeologist's own body (Barthes again: 'revealed, riven, exposed, in the photographic sense of the term' [Barthes: 1977: 34][9]); this is an effigy that calibrates time. Has the body aged twice, once outdoors in flesh and once indoors in paper? Gerhard Richter once suggested that 'What the photograph mourns is both death and survival, disappearance and living-on, erasure from and inscription in the archive of its technically mediated memory' (Richter 2010: xxxii). Perhaps our/my autopoietic bodies might also identify with the deferral between the production of a negative and the printing of an image. This image (human, environmental, Roman, 1990s, anticipatory) approaches us, unrepressed, from the future and the past at once. From the central reservation of the A46 travelling north from Evesham. These two bodies now re-encounter one another in this once future moment, in a 'fortuitous meeting of two distant realities on an inappropriate plane' (Ernst 1970 [1936]: 256).[10] One part a deferral from the past, like the square bracket through which we index translations ('This is 1967,' writes Deleuze in English translation [2004 (1967): 170]); two parts the decomposition of myself.

Translator Kate Briggs has observed how, in concluding his inaugural lecture at the Collège de France, Roland Barthes made some connections between 'three moments of a disease', reflecting on the account of tuberculosis in Thomas Mann's *Der Zauberberg* (published 1924, set in 1907) through Barthes' own experience of suffering from tuberculosis in 1942, and his (then) present moment in the lecture theatre in Paris on 7 January 1977, by which time the disease had come to be effectively 'vanquished by chemotherapy'. Barthes is 'stupefied' by the realization that the first two moments, a moment of his past reality and a moment of fiction set eight years before his birth, could 'merge, far from his own present', could be contemporary with each other, because the strong connection between the virulence of the disease before its treatability made these moments 'equally remote' – the realization that '*my own body was historical*', that 'my body is much older than I,' then in order to live he must forget:

> I must fling myself into the illusion that I am contemporary with the young bodies before me, and not with my own body, my past body.
>
> BARTHES 1978: 15–16[11]

This is not quite Lacan's notion of *aplianisis* (the fading of the subject) that Barthes had employed the previous autumn in his James Lecture, on the theme of *Proust et Moi*,[12] but a kind of prefigurative distribution of personhood. It operates in the only register through which fiction can transform into reality: the register of temporal imagination. Susan Sontag's confident distinction between archaeology as 'inventory' (along with weather forecasting and espionage) and fiction breaks down.[13] Briggs explains that the notes for the lecture course at the Collège de France that Barthes began on 12 January 1977, a week after this inaugural, were not published in French until 2003, and that her own English translation came a full decade later:

> These lags in publishing and translating that produce new readerships: bodies like my own, as yet unborn at the time of the lectures themselves listening now to the sound files of the audio recordings, reading the notes, making them speak and be spoken to by – making them contemporary with – my own present moment. "Who are my contemporaries?" Barthes would ask in a lecture given a few months later: "Whom do I live with?" The calendar, telling only of the forward march of time, is of little help. The way it brackets together work produced in the same set of years, as if shared historical context were the condition or the guarantor of a relationship. The way it holds more distantly dated relations apart. My copy of *The Magic Mountain* lies open next to Howard's translation of the inaugural lecture, the one that was delivered before a packed auditorium; all those young bodies, they must be older now, pressed together in their seats, the aisles, out onto the corridors.
>
> BRIGGS 2017: 17

These confusions of contemporaneities reveal the untimely nature of photological knowledge (cf. Hicks 2003). As the eyepiece of an astronomer's telescope is lit up by the most remote past refracted across light years, so the experience of this archaeographic encounter seems like an intervention in the passage of time that casts a light on my skin, a projection like the projections of that magic lantern in Proust. Through the 'lyrical resurrection of past bodies' we might 'make History into an enormous anthropology' (Barthes 1978)[14] and reveal 'the illusory logic of biography' if it follows a 'purely mathematical order of years', since 'systems of moments succeed each other, but also respond to each other' (Barthes 1984 [1978]: 318).[15] My hands returned the body that had laid on the museum lightbox to the embalming grip of its plastic folder, where it remains suspended from a metal strip under lock and key in a fireproof cabinet. What I share with you here is a mutual knowledge made through two corporeal forms of my own temporal thinking that have been living apart and now make recollection a possibility. 'The translation from idea or effort into material object and back again marks the boundaries of the person,' suggests Marilyn Strathern (2005: 154).[16]

What I am describing to you is both my own body and is not. It (this image will not freight the first person singular) does not index my past. We (I look through this screen, fingers on keys at the archaeologist with eyes downcast focused on the ground) and I (not a trace or a ruin but an untimely ghost returning from the future to haunt this photograph) resist any variety of reflexivity that would offer to this two-dimensional form a second biography. It is by no means just some sense of the uncanny ('that small terror that is there in all photography: the return of the dead' [Barthes 1980: 23][17]). I/we resist being reduced to a construction or a metaphor. Me, us and it emerge not through analogy (which would be a confusion of value) but through these confusions of practice: *we appear to each other*. Should I believe that this is a 'representation of a possible world and not as a necessarily real having-been-there' (Dubois 2016: 161)? Surely some kind of trace still, at least, even if what is indexed is somehow speculative? Not quite – this glimpse of a now transformed landscape, the four-lane highway from the ploughed fields of Alcester to the orchards of Evesham, looks through my body, as if an archaeological image could somehow, like a police surveillance camera, be body-worn. As a method, Visual Archaeology has implicated my body in a transformation of the past. I *identify with* this knowledge, just as excavation, it is an experiment performed with one's own body. And what is method after all if it is not 'a premeditated decision' (Barthes 2013: 3)?

*

VIII. Unhistoricity

The affective, even psychoanalytical effects of looking back through the ongoing presence of the image as if it were something other than a trace operate as a visual methodology. Through this mnemic presence the apparent suddenness of the snapshot, that most 'abrupt artefact' (de Duve 1978: 117), is slowed with a stillness that identifies with the archaeological object. As light is refracted backwards, regresses towards the past through the image as *Erinnerungsspur*, we might almost believe in Pierre Bourdieu's suggestion that

> Photography has the function of helping one to overcome the sorrow of the passing of time, either by providing a magical substitute for what time has destroyed, or by making up for the failures of memory, acting as a mooring for the evocation of associated memories, in short, by providing a sense of the conquest of time as a destructive power.
>
> BOURDIEU 1990: 14

But in the case of the knowledge of the past that emerges through archaeology's visualism, things are too unstable for such moorings or conquests. Photology displaces all trace of melancholy, in a manner akin to what David Berliner calls anthropology's *exonostalgia* 'which encompasses discourses about loss detached from the direct experience of losing something personal' (Berliner 2014: 376). What is 'the ontology of the photographic image?' asked André Bazin six decades ago:

> If the plastic arts were put under psychoanalysis, the practice of embalming the dead might turn out to be a fundamental factor in their creation. The process might reveal that at the origin of painting and sculpture there lies a mummy complex. The religion of ancient Egypt, aimed against death, saw survival depending on the continued existence of the corporeal body. Thus, by providing a defense against the passage of time it satisfied a basic psychological need in man, for death is but the victory of time. To preserve, artificially, his bodily appearance is to snatch it from the flow of time, to stow it away neatly, so to speak, in the hold of life. It was natural, therefore, to keep up appearances in the face of the reality of death by preserving flesh and bone. The first Egyptian statue, then, was a mummy, tanned and petrified in sodium. But pyramids and labyrinthine corridors offered no certain guarantee against ultimate pillage. . . . No one believes any longer in the ontological identity of model and image, but all are agreed that the image helps us to remember the subject and to preserve him from a second spiritual death. Today the making of images no longer shares an anthropocentric, utilitarian purpose. It is no longer a question of survival after death, but of a larger concept, the creation of an ideal world in the likeness of the real, with its own temporal destiny.
>
> BAZIN 1960: 4–6

But there is more at stake in photological knowledge than some ontological multiplication of personhood; the temporal effects of the interruption of change involve more than preservation and a new history or biography for the image. Consider the immense rapidity with which archaeology pioneered the use of radiography. Made less than two years after Röntgen's discovery of the new imaging technique, Flinders-Petrie's X-Ray of wrapped mummified leg bones dating from the Fifth Dynasty, excavated from the site of Deshasheh in 1897, comprised tibias, fibulae, tarsals and metatarsals from perhaps three individuals (Petrie and Griffith 1898; Figure 12.3). Petrie's published image offered no commentary at all on his presentation of the result of the slow exposure of the glass plates of the X-Ray of a 4,500-year-old mummification. As one of first publications of X-Rays in archaeology, it came shortly after radiographs of mummified human and animal remains made by Walter König in Frankfurt and Thurston Holland in Liverpool (Adams 2016: 371), just as a new technique for unwrapping a mummy with the minimum of physical intervention was developed. In this 1898 radiographic 'unwrapping' we can clearly discern Harris Lines on the tibias – a growth arrest phenomenon made visible through radiography but

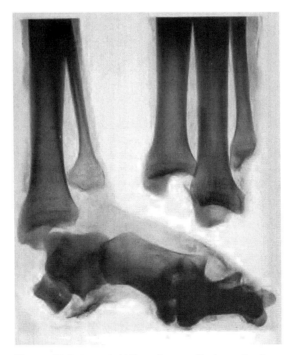

Figure 12.3 An early X-Ray of mummified remains from Deshasheh, Egypt published by William Flinders Petrie and Francis Llewellyn Griffith in 1898 with the caption 'Deshasheh. X Ray View of Legs of Dissevered Bodies in Wrappings' (Petrie and Griffith 1898, Plate XXXVII).

which was unknown until identified twenty-nine years after this image was made (Harris 1926).[18]

These untimely Harris Lines reveal the temporal qualities of photological knowledge, in that they present a problem that is the inverse of survival: in the moment of photographic exposure they are made visible but cannot yet be seen, while we can clearly identify them today. In a well-known attempt to tackle such questions, Bruno Latour pushed beyond constructivist accounts of scientific knowledge into the territory of what he called 'the historicity of things', exploring the effects of scientific discovery by asking, 'Where were microbes before Pasteur?' (Latour 1999a: 145). Calling for some way to think outside conventional distinctions between 'emic' and 'etic' in accounts of the history of ideas on the one hand and the history of science on the other, Latour sought to explain how the scientific object (lactic acid ferments) was not just 'waiting for the light to fall on it', but required an 'industrial' as well as an 'optical' metaphor to describe how it was 'made visible' through the discovery by Pasteur in 1864 (Latour 1999a: 139–40). To complement the history of discovery and visibility, Latour introduced a second temporal axis using the archaeological metaphor of 'the sedimentation of time' to express the ongoing history of the retrospective 'production' of the past. 'After 1864 airborne germs were there all along,' he suggested, since 'the year 1864 that was built *after* 1864 did not have the same components, textures, and associations as the year 1864 produced *during* 1864',

> What does it mean to say that there were microbes 'before' Pasteur? Contrary to the first impression, there is no deep metaphysical mystery in this long time 'before' Pasteur, but only a very simple optical illusion that disappears as soon as the work of extending existence *in time* is documented as empirically as its extension *in space.* My solution, in other words, is to historicize more, not less. No sooner had Pasteur stabilized his theory of germs carried by the air than he reinterpreted the practices of the past in a new light . . . Pasteur *reinterpreted* the past practices of fermentation as fumbling around in the dark with entities against which one could now protect oneself.
>
> LATOUR 1999a: 169

In Latour's account, this doubling of temporal axes through the idea of moments of time as built in retrospect through changing present networks of people and things meant that there was not only 'the year 1864 of 1864', but also 'the year 1864 of 1865', 'the year 1864 of 2008', etc. (Latour 1999a: 170, 172):

> In this second dimension there is also a portion of what happened in 1864 that is produced *after* 1864 and made retrospectively a part of the ensemble that forms, from then on, the sum of what happened in the year 1864.
>
> LATOUR 1999a: 172

Latour contrasted the 'linear succession' of the optical axis of discoveries and revelations with the 'sedimentary succession' of making and re-making. But this sedimentation is also unilinear in accumulation, albeit retrospective and changing in reinterpretation. Thus, the year 1864 may be retrospectively reconstituted or 'retrofitted' in successive stages, but these stages themselves are subject to the 'irreversible movement of time' that only ever moves forwards (Latour 1999a: 171), in a manner markedly similar to the Representational Archaeology's account of reconstructing the past through its traces. Strangely, Latour appears to have the 'sedimentation' axis upside-down, since surely no kind of successive sedimentation can deposit the present stratigraphically earlier than the past. This is a mistake that is not present in the opening lines of Foucault's *L'Archéologie du Savoir* which is where Latour's sedimentary/succession distinction appears to derive ('[New] tools have enabled [us] to distinguish in the field of history various sedimentary strata; linear successions, which for so long had been the object of research, have given way to discoveries in depth' (Foucault 1969: 9–10)[19]), but which Latour himself reinforces in his commentary:

Substance does not mean that there is a durable and ahistorical 'substrate' behind the attributes, but that it is possible, because of the sedimentation of time, to turn a new entity into what lies *beneath other entities*.

LATOUR 1999a: 170, original emphasis

Although Latour impossibly inverted the axis of sedimentation, so that the contemporary continually undercuts the layered past, both of his axes, seeing/discovering and making/constructing, became vectors: unilinear and progressive. And yet for the archaeological image, this temporal framework must be more complex. Alongside the methodological question of discovery (Harris Lines, 1927) and retrospect (our ongoing understanding of them ever since, and into the present), there are also the practical questions of the extension of images across the deeper past: this X-Ray (1898), this discovery (1897), this mummy (Fifth Dynasty, 25th–24th centuries BCE) and those human/cultural and material/natural taphonomic processes (what Mike Schiffer [1975] called 'n-transforms' and 'c-transforms') that interdigitate between them.

Let us disrupt, or at the very least transform, the optimistic superposition of knowledge and light that characterizes Latour's super-historical account of scientific knowledge, through a consideration not of constructed facts but of inadvertent fictions. Just as the so-called 'weak programme' of the sociology of scientific knowledge used the failure of technologies or theories – 'error, irrationality and deviation' (Bloor 1999: 81) – as heuristic opportunities, so a

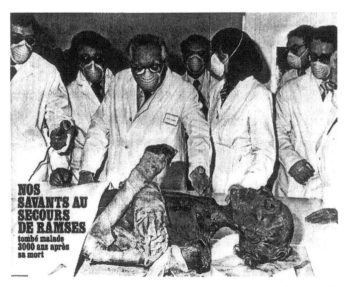

Figure 12.4 'Nos savants au secours de Ramses, tombe malade 3000 ans après sa mort.' *Paris-Match* 26 November 1976 (Tardrew 1976).

mistaken account of the past (offered as an example in the spirit of identification with, rather than comparison to) might help us to interrupt the forward-march of what Latour imagines to be two irreversible axes of onward-moving chronology. Let us suggest that the tick of the camera and the tick of the artefact (recall Barthes: 'the living sound of the wood') might turn out after all to be different, in both character and effect, to the tick of a clock. Like Petrie's X-Ray, although more than a century later in date, the account in question concerns Ancient Egypt.

'Something is at stake here that defies the normal flow of time,' suggests Latour. He is questioning how a body, three millennia after its death, could be 'endowed with a new feature' (Latour 1999b: 247, 249). The body at hand lies on the page of a 1976 *Paris-Match* report which, Latour claimed, suggests that Ramesses II (d. 1213 BCE) died of *tuberculosis bacillus* – first identified (or discovered? or described?) by Robert Koch in 1882. Whatever our preferred verb, the headline is unchanging: '*Nos savants au secours de Ramses II tombe malade 3000 ans apres sa mort!*' (Figure 12.4). 'The attribution of tuberculosis and Koch's bacillus to Ramses II,' Latour explained,

> should strike us as an anachronism of the same caliber as if we had diagnosed his death as having been caused by a Marxist upheaval, or a machine gun, or a Wall Street crash. Is it not an extreme case of "whiggish" history, transplanting into the past the hidden or potential existence of the future? . . . We are faced with a new ontological puzzle: the thorough historicization not only of the *discovery* of objects, but of those objects *themselves*.
>
> LATOUR 1999b: 248

Latour's argument was concerned here with post-mortem diagnosis and scientific fact in terms that are wider than epistemological questions, understanding 'historicity' as more than just a human preserve (Latour 1999b: 250, 263–5). He sought to extend history to scientific objects as well as just scientific knowledge, an idea that he subsequently developed through the notion of 'the double historicity of science and of its subject matter' (Latour 2007: 2).

But at stake in this instance is not scientific method and historical fact – but sociological fiction. In a footnote, with a reference to Desroches-Noblecourt's (1996) book about Ramesses II, Latour states that 'In spite of the flippant titles usual for *Paris-Match*, a reading of the text shows that it is not actually the king who has become sick after his death, but rather the mummy, from an infection by a fungus' (Latour 1999b: 247, note 1). In truth, the headline refers not to a retrospectively discovered disease. Returning to the original 1976 *Paris-Match* text, we find the journalist Catherine Tardrew's description of the politics of conservation and decay:

It is the final political adventure that the powerful pharaoh of ancient Egypt will have known: Rameses II, who had seen the exodus of the Hebrews, suffered the invasion of the locusts and defeated the fierce Hittites, is today at the centre of a bitter polemic between Anglo-Saxon and French academics. The mummy of the great king was welcomed at Le Bourget airport two months ago with the kind of honours reserved for heads of state: a minister, the republican guard, and a military band. The purpose of the visit: to stop the decomposition of the embalmed corpse that had been documented in the Cairo Museum. '"It is a diplomatic disease", insinuate some English and American Egyptologists. "Our French colleagues are justifying the arrival of the pharaoh to Paris with this supposed damage."' Professor Lionel Balout, Director of the Musée de l'Homme and responsible for Operation Ramses, is indignant: 'Absurd. The smell of the mummy in the Cairo Museum alone would prove the state of decomposition. The bands had been removed by looters of the royal tombs and the body was barely protected by linen put in place centuries after the embalming. We discovered microscopic fungi, bacteria and even larvae of recent origin. Another tragedy: tensions are occurring between the flesh dried out by the resins and the skeleton, causing major fractures.' To stop this from happening, the body will be radiation-sterilized with cobalt-60. The treatment of Ramses II is scheduled to end in March. The Egyptians will then be reunited with their dead sovereign, the first that they have allowed to travel to a foreign land.

TARDREW 1976: 76[20]

'I nonetheless have kept the first interpretation, associated with the image, because of its ontological interest,' Latour insisted (1999b). The philosopher's misdiagnosis was based on a misreading of the newspaper snapshot and headline.[21] But there is simply no evidence for Latour's account of TB in this mummy. The 'ontological interest' has no substance: it is nothing but a misinterpretation. We must therefore agree with Latour's answer to his own question 'Did Ramesses II die of tuberculosis?' 'No!' – but not because it would pre-date the discovery of Koch's bacillus by 3,075 years, but because there is no suggestion, from archaeological evidence or journalistic reporting, that the mummy of Ramesses II showed any signs of TB – *Paris-Match* did not even make this claim. (The rather bulky ossification has been diagnosed by Zahi Hawass and Sahar Saleem (2016: 162) as diffuse idiopathic skeletal hyperostosis (DISH), also known as Forestier disease (Forestier and Rotes-Querol 1950) – a common condition among elderly people characterized by abnormal calcification or bone formation [*hyperostosis*] around the spine.[22]) At stake in the *Paris-Match* story was not the philosophical idea of diseases existing before they are diagnosed, which is a question of the interpretation of historical context. *Latour mistook decay for disease.* The archaeology of decay rather than of disease which, like the archaeological photograph, like archaeological knowledge, constitutes not a frozen moment of the trace or fragment, but an ongoing process – the decomposition of historical contexts.

How then to comprehend the Harris Lines that we see in Petrie's X-Ray? Let us suggest, against Latour, that photological knowledge operates across time in a manner that his account of forward-marching and retro-fitting does not accommodate, in that transformation can operate in more than a unilinear, progressive manner. Retrospection can constitute more (or perhaps, less) than just reinterpretation. Key here is how to underdetermine the intentionality of the photographer as interpreter in the first place. We need to question the level of visual and temporal control (a remnant of hyper-Foucauldian thinking, surely?) that we often hold as implicit, as for example in Tim Ingold's account, following Berger, of how 'the photograph arrests time' whereas in contrast 'the drawing flows with it': 'the still camera arrests a moment in both the consciousness of the photographer and the things that hold his attention, and effects an instantaneous capture, of the latter by the former' (Ingold 2013: 127, 231).

These oracular Harris Lines within an X-Ray taken some three decades before their discovery surely annihilate the idea of irreversibility that lies at the heart of Latour's 'historicities' and Ingold's 'arrests', and of any idea that Photography, and thus Photology, might somehow halt time – even were it to employ such violence as that to which Barthes refers in various places in his writing to connect the closure of the shutter and a bringing about of death. Far from any sense of a trace, through which the image or knowledge persists 'to defy the ordinary wear and tear of time' (Henderson 1869: 37), instead there is in the photological some kind of, as Fox Talbot expressed it right at the start of Photography's history in *The Pencil of Nature*, a deferral or hesitation not as a loss of control or a failure of intention but as a knowing effect inherent to the photographic image as knowledge of the world:

> The operator himself discovers on examination, perhaps long afterwards, that he had depicted many things that he had no notion of at the time. Sometimes inscriptions and dates are found upon the buildings, or printed placards most irrelevant, are discovered upon their walls: sometimes a distant dial-plate is seen, and upon it – unconsciously recorded – the hour of the day at which the view was taken.
>
> FOX TALBOT 1844: 40

In this territory of the unexpected revelations of photology as a practice, we encounter not the chance fragment as a trace, but something closer to Barthes' account of the 'reality effect' of the inclusion of insignificant detail in a text – 'language fading into the background, to be supplanted by a certainty of reality: language turning in on itself, burying itself and disappearing, laying bare what it says' (Barthes 2013: 70). Through such details, Barthes suggests, fiction establishes authenticity. We might recall in this respect how Julia Margaret Cameron 'availed herself of badly made lenses in order to get at the "spirit" of the person portrayed without the disturbing interference of "accidental detail"' (Kracauer 1960: 6–7). As the invention of writing meant that societies could 'amass knowledge . . . by keeping records of events, and storing up new observations for the use of future generations' (Tylor 1881: 179), so the temporal jumbling that comes with the territory of photological knowledge constitutes more than the accidental capture of unintended detail, which has been the focus of what Visual Anthropology, resurrecting an idea from film theory, has come to describe as the 'pro-filmic' – the idea of an unanticipated or accidental superfluity of detail captured in the 'micro-event' of the shutter closure through which the artifice of the photographer breaks down or is surpassed (Pinney 2005). The example of a misrepresentation – a tuberculosis that was philosophically misdiagnosed, decay which was sociologically mistaken for a disease – shows how photological knowledge does more than re-interpret in retrospect: it operates outside of the supposedly 'unidirectional successions' of knowledge, whether 'linear' or 'sedimentary', in an unhistorical manner: to correct a mistake, or to change the past in some other manner. Any unintended detail, not just a mistake, can thus through the photological involve causation operating from present to past as much as past to present.

Our encounter through the newspaper image with Bruno Latour's mistaken account of the tuberculosis of Ramesses II (which we know did not exist) in Paris in 1976 thus operates to some extent like Hans Castorp's visit to his tubercular cousin, Joachim Ziemssen, in the sanatorium in Davos in 1907 (in that it is a fiction). But what we learn from this is that the same goes for any nonfictional photological past, like the Harris Lines present in Petrie's X-Ray in 1897 – if we see them photologically, then we understand these images as transformations of time, and thus not just archaeologically but unhistorically.

*

IX. Impermanence

This photograph is not still (Figure 12.5). Two crayons bound together found on 6 December 2016 dropped by the side of the service road at the informal 'Jungle' refugee camp in Calais, a few days after its closure. A child's object dropped or discarded, the rubber band a gesture that improvises an ordering or collecting under conditions of relentless displacement, an archaeology of hospitality, of parenthood, of loss, of conflict, of protest, of precarity, of the European border regime (Hicks and Mallet 2019). 'Photographs are retrospective' asserted Berger (Berger and Mohr 2016 [1982]: 279), but to photograph impermanence is to look

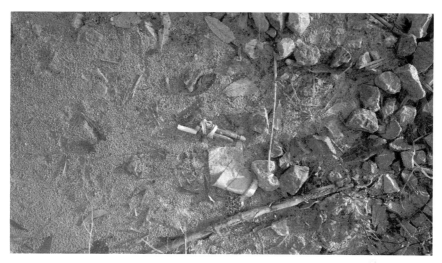

Figure 12.5 Two crayons bound with an elastic band, dropped at the side of a service road at the Calais 'Jungle', 6 December 2016 (photo: Dan Hicks)

back before the dust has settled – and thereby to bear witness to the classification of other humans to that most abject of human categories, the archaeological. Here the temporal intervention of Visual Archaeology is with objects, places and human lives in which questions of time, transience and endurance are not just abstract conditions of existence but tools for the governance of displaced populations. A photological gesture could begin a process that might take the form of some kind of counter-archive, witnessing undocumented lives – in Warburgian terms survival that is transformed through sight towards Nachleben, the possibility of 'a time other than habitual chronologies' (Didi-Huberman 2002: 61). This snapshot can hardly be said to represent some kind of formal archive, but the visual knowledge towards which it points might offer some kind of counterpoint to precarity, not some simple abrogation of transience but a kind of contemporary hindsight in the mode of *anastrophe* (the reversal of the order of words), or *antistrophe* (the turning of an argument against itself). This is not a question about photography as a medium or a practice, or the existential questions asked by Sabine Kriebel (2007: 3): 'What is a photograph? What is photography?' The archaeologist asks: *What has gone before, and what will come next?* The image asks: *When is this photograph?* Lévi-Strauss (1997 [1993]: 30) observed how: 'Sometimes beauty lasts only for an instant. As the word snapshot suggests, photography seizes this moment and reveals it.'[23] Horror, loss, abjection too can be fleeting, can pass, but for displaced people in Europe that very fleetingness represents a form of ongoing governance. In this image, the photological knowledge in the flash of revelation described by Lévi-Strauss is not instantaneous but distended, episodic in form. It is not a frozen moment in time but an ongoing event distributed across time and space in the same way as Lévi-Strauss suggested that archives embody *événementialité*, they 'constitute the event in its radical contingency' while also 'giving a physical existence to history that overcomes the contradiction between a bygone past and a present where it survives' (Lévi-Strauss 1962: 321).[24]

As neither primary evidence nor secondary scientific documentation, but 'the ontologies created and sustained by scientific observation' (Daston 2008: 97) that constitute a kind of 'third nature' (Daston 2017), photological knowledge can reveal undocumented lives even in the contemporary world. In the refugee camp, impermanence is a principal form of governance. The archaeological photograph of this impermanent near-present 'does not distinguish accidents from the nature of things, but places them both on the same level' (Lévi-Strauss 1997 [1993]: 29).[25] In Calais the instantaneous, in the form of the emergency and impermanence, is a neo-colonial method for the classification of human bodies (as illegal or legal; as objects or subjects). The work of photology here, knowledge of lives that are brought into sight, is a resource for visual advocacy, grounded in its nonverbal productions that seeks to 'read what was never written' (Benjamin 1974 [1940]: 1238).[26] Re-inscribing human life, the image recalibrates time with its focal length, compressing into the frame dense fragments from the very recent past that are transformed through trauma, loss, conflict, violence. Photological documentation intervenes through the imagination of human life, focused, in Barthesian terms, not on the photographic *noeme* '*ça-a-été*', but the 'as if' (*comme si*), not fixed on the remnant but at once both scientific and imaginary (compare Barthes 1980: 120, and 1984 [1978]: 325).

The exposure of the contemporary brings about, just like the Visual Archaeology of any more distant past, a transformation of time that does not extend human biographies through an image but redraws the scale of events, indefinitely or recursively, through the photological technology of the archive (cf. Hicks 2010). Barthes describes it as 'an anterior instant':

> The image is perfectly adapted to this temporal decoy: clear, surprising, framed, it is already (always already) a memory (the nature of the photograph is not to represent but to remember).
>
> BARTHES: 1977: 229[27]

Like the repeatedly photocopied photographs, heavy on the toner, that became some grainy part of language for W.G. Sebald that eroded the indexical character of the relation of image to object, transforming the blurred image into a kind of speech act (Harris 2001; cf. Hicks 2016: 6), so photological knowledge of the present day is a transformation and an accident. A host of precursors include the works of André Breton, the found images that constituted moments of chance (*hazard*) in the Parisian streets of *Nadja* (Breton 1928). Like the photographic

(Kelsey 2015: 1), the archaeographic is surely 'prone to chance'; more than it is prone to ruin. What Silverman (2015: 33) describes as 'the second coming of the world' involves in photological knowledge of the present a turning-around that interrupts any representation of the past, as *analepsis* (flashback), *what-has-been*, gives way to *peripeteia* (a sudden reversal) and anagnoritic knowledge (discovery): as if it were *what-appears-to have-been*.

Photological knowledge of the present (part of what is often called 'contemporary archaeology') reveals a much wider problem with how Photography has often been thought of in relation to the instant. Like so many theories informed by the literature of the *photographique*, the Representational Archaeology was dominated, through the rhetoric of the trace, with a particular account of the momentary. As we have seen, its artefactualist theory of the remnant drew liberally on the idea of the photograph as a frozen moment of time, a mummification of the world, overly influenced perhaps by the marketing of the instant camera in the 1970s and 1980s, the Polaroid as the embodiment of the idea of the instantaneousness of the camera, the brand names (Presto!, Pronto!, Timezero Onestep), that strange materiality of the disposable, one-time-use Polaroid Flashbar or the Kodak Flashcube, where ten flashes of light corresponded with the ten exposures of the filmpack, not just producing the photograph with a whirr but leaving behind a line of clouded bulbs.

An alternative sense of the instant might begin with Duchamp's account of the 'snapshot effect', which requires 'planning for a moment to come', a 'matter of timing' creating 'a kind of rendezvous' (Duchamp 1934). We might read this alongside André Breton's account of *hazard objectif* (objective chance) which brings about 'the intersection of subjective and objective lines of development' between internal thoughts and external realities in 'determined coincidences' (Cardinal 1986: 21), a theory of the forging of connections (Caws 1988: 93). And we might recall the centrality of this theory of Breton's to Lévi-Strauss's account of the effects of the classificatory work of the bricoleur:

> Once realised it will inevitably be out of step with the initial intention (which was simply a schema), an effect that the surrealists have called 'objective chance'.
> LÉVI-STRAUSS 1962: 31–2[28]

The photograph is not a still. It's 'not yet' in the sense of *Noch-Nicht-Bewusste*, and so only partly like a flashlight shone into what Ernst Bloch called *das Dunkel des gelebten Augenblicks* ('the darkness of the lived moment') (Bloch 1959), in that it could never fully represent one particular, primordial human blink of the eye to be reinterpreted in retrospect. It sets off a provisional movement that is the obverse of the conflict, forced migration and grinding human impermanence that led to this moment of deposition. The idea of the standstill was in any case only created through the modernist possibility of imagining the unmoving image made possible by cinema (Hornby 2017). Photological knowledge of the contemporary diffuses the luminosity of the Benjaminian flash of memory, recasts available light, reverses the Proustian magic lantern effect, reframes the Barthesian punctum as an experience of loss, 'that accident which pricks me (but also bruises me, is poignant to me)' (Barthes 1980: 49),[29] by salvaging duration through the documentation of abjection, precarity, instability. Photology is visual knowledge that holds 'a power of expansion' (Barthes 1980: 74).[30]

*

X. Appearance

The Victorian image of the soldier-archaeologist is transformed into a mask. 'Othering the Ethnologist, Augustus Pitt-Rivers' is a photographic work by Australian Aboriginal artist Christian Thompson. It forms part of the series 'Museum of Others' and uses a historic photographic portrait of archaeologist General Augustus Henry Lane Fox Pitt-Rivers (Figure 12.6) from the collections

Figure 12.6 Photographic portrait of Augustus Pitt-Rivers, probably dating from 1886 (Pitt Rivers Museum 1998.356.80).

Figure 12.7 Christian Thompson 'Othering the Ethnologist, Augustus Pitt Rivers' (Museum of Others series). C-type print on metallic paper, 2016 (Pitt Rivers Museum 2017.63.1).

of the Pitt Rivers Museum, which the General founded at the University of Oxford in 1884, to produce a new 'conceptual self-portrait'. A second photographic portrait is made (Figure 12.7), which excavates the self-image of the British collector, ethnologist, and archaeologist through Thompson's Indigenous (Bidjara) identity. One carefully orchestrated pose is transposed into another, from London via Oxford to Central West Queensland, addressing questions of cultural heritage and colonial entanglement through the mode of self-portraiture, working against the legacies of the representation of Aboriginal people by Western anthropologists who assigned an archaeological, prehistoric identity to living communities.

Thompson, as Jane Lydon has put it, is 'transmuting' Aboriginal photography, to 'displace the historical markers of identity central to colonial photography – especially the anthropometric mugshot' (Lydon 2014: 13, 2016). In doing so he reimagines the conceptions of time that lay behind the representation of Indigenous people – evolutionary, degenerative, the idea of the primitive, survival, the fantasy of the Stone Age in the contemporary world. The photograph operates within this history of science, of subjugation, of land-grab, of colonial ideologies of race that destroyed Indigenous communities, of violence. ('The recollection of death . . . is part and parcel of every memory-image', Kracauer

1993: 433.) Thompson challenges the authority of the museum as 'a burial chamber of the past',

> to perform a 'spiritual repatriation' rather than a physical one, fragment the historical narrative, and traverse time and place to establish a new realm in the cosmos, set something free, allow it to embody the past and be intrinsically connected to the present.
>
> THOMPSON 2012: n.p.

The image contravenes any protocol of the avoidance of looking at images of the deceased. It uses the Indigenous concept of 'the Dreaming' or 'Everywhen' not to assert the contemporaneity of the historic image, but to express the inseparability of past and present (Gilchrist 2016: 23), and not to reduce the artistic intervention to research, or to imagine the artwork as a contact zone between pre-existing cultures, disciplines or periods of time. Rather it begins a transformative and improvised 'democratization' that reverses the colonial gaze to throw light on relics, silences and unseen pasts of the archive through new engagements of relatives and descendants of the images' subjects (Lydon 2016) – a new layer in the history of how as a Western technology photography has been appropriated locally as a medium around the world, disturbing and transforming Western culture (Pinney 2003: 1, and 2010: 149). ('The passage of time,' as Genette (1980: 156) remarked in his discussion of Proust, can be 'masked behind repetition.') The process of witnessing the transformation of colonial history that Thompson effects through his photological cosmopolitics, through a visual politics of unforgetting. We might compare this with Barthes' account of the living and the historical in photography:

> As a living soul, I am the very opposite of history, one who denies it, destroys it for the sake of my own history (impossible for me to believe in 'witnesses'; or at least impossible to be one).
>
> BARTHES 1980: 102[31]

In Thompson's autoethnographic and emancipatory photographic work the mask does not straightforwardly represent Indigenous identity but uses the resources of Archaeology and Anthropology to reverse and refract the colonial gaze, inverting the exclusions of exoticization and othering in something close to what Lévi-Strauss described as a transformation:

> A mask is not first and foremost what it represents but what it transforms – what it chooses not to represent. Like a myth, a mask denies as much as it affirms; it is made not only of what it says or what it understands itself to say, but also of what it excludes.
>
> LÉVI-STRAUSS 2008 [1979]: 978[32]

Augustus Lane Fox, before he adopted the name Pitt-Rivers, before his photographic portrait, and before Thompson's mask, suggested that archaeology should study words and tools in the same manner:

> Words are the outward signs of ideas in the mind, and this is also the case with tools or weapons. Words are ideas expressed by sounds, whilst tools are ideas expressed by hands; and unless it can be shown that there are distinct processes in the mind for language and for the arts they must be classed together.
>
> LANE FOX 1875: 500

The human hand as a tool or a weapon surely precedes the stone axe. But perhaps the primary human tool, the first body part capable of transforming the world, was not the hand, but the eye. After all, the photograph does not (*pace* Barthes 1961) reduce the object to its image; and the reverse process, through which anthropological material culture studies might reduce the image to an object, would equally fail to account for the kind of visual knowledge and transformation that we witness in Thompson's work. What vocabulary to use for Thompson's photological intervention, his transformation of the visual knowledge made through Archaeology and Anthropology that destabilizes the present, a transformation through looking? Thompson's image does not make mute fragments of the past speak for untold histories, offering alternative views upon on the past, but collapses cosmopolitics, biopolitics, necropolitics into the temporal uncertainties of photological knowledge.

We might call these transformations 'appearances', to capture how we might mistake them for mere superficialities that mask some deeper more stable reality, how they emerge through practice, how they intersect the alternate invisibility and hyper-visibility of the silencing and surveillance of non-white, non-Western lives. Doing so would be to draw from how Nick Mirzoeff has reformulated Hannah Arendt's notion of 'the space of appearance' in his account of Black Lives Matter: photography of police violence and killings, protest and social media from 'cell-phone videos and photographs, supplemented by machine-generated imagery taken by body cameras, dash cams, and closed-circuit television footage'

> I will call the interface of what was done and what was seen and how it was described as 'appearance', especially as the space of appearance, where you and I can appear to each other and create a politics. What is to appear? It is first to claim the right to exist, to own one's body, as campaigns from antislavery to reproductive rights have insisted, and are now being taken forward by debates over gender and sexual identity. To appear is to matter, in the sense of Black Lives Matter, to be grievable, to be a person that counts for something. And it is to claim the right to look, in the sense that I see you

and you see me, and together we decide what there is to say as a result. It's about seeing what there is to be seen, in defiance of the police who say 'move on, there's nothing to see here', and then giving the visible a sayable name. People inevitably appear to each other unevenly – the social movement process is about finding ways for people to learn how to treat each other equally in circumstances where they are not equal, whether in material terms, or those of relative privilege. To take the foundational example, the indigenous person in the Americas always knows that the land in which we appear was stolen from them and so the work of creating the space of appearance is always decolonial.

MIRZOEFF 2017: 17–18

Here are some political dimensions of photology – the kind of politics that Jacques Rancière described as a 'demonstration that makes visible that which had no reasons to be seen' (Rancière 1997: 101)[33] – that have begun to be explored in a number of recent studies, most notably Ariella Azoulay's (2012) critique through the lens of contemporary Palestine of the focus in some of Barthes' work on photography as a form of representation, in the place of which the idea of *Civil Imagination* introduces the notion of the photograph as an unfinished event (Azoulay 2012: 25). In a similar vein, Nick Shepherd's *Mirror in the Ground* (2015) considers the politics of the photographic representation of 'native labour' on archaeological excavations, and Colin Sterling's account of tourist photography at Angkor Wat and other iconic heritage sites shows how the mass-production of images can cause sites themselves to be actively transformed and discusses various acknowledgements and negations of 'the "aftershocks" of colonial photography' (Sterling 2015: 145; cf. Sterling 2017). (Barthes: 'It's not just that a photo is never, in essence, memory (whose grammatical aspect would be the perfect tense, whereas the tense of the photo is the aorist) but that it blocks it, quickly becoming a counter-memory.' [Barthes 1980: 142][34])

And yet few such studies have come close to addressing the temporal politics of cultural appropriation that are so often encountered in early anthropological accounts of photography, as for example in Lévy-Bruhl's (1922) account of Junod's discussion of photography and the parting of body and soul among the Thonga people in South Africa:

Almost everywhere photographic equipment appeared especially dangerous. 'Ignorant natives', says Junod, 'instinctively object to being photographed. They say: "These white people want to rob us and take us with them, far into lands which we do not know, and we shall remain only an incomplete being." When shown the magic lantern you hear them pitying the men shown on the pictures and saying: "This is the way they are ill-treating us when they take our photographs!"' Before the 1894 war broke out, I had gone to show the magic lantern in remote heathen villages. People blamed me for causing this misfortune by bringing back to life men who had died long ago.

LÉVY-BRUHL 1922: 440[35]

How to begin to acknowledge how much was 'robbed' through the anthropological camera in this way? Thompson's artwork certainly constitutes not only a 'modification', in the Husserlian vocabulary used by Alfred Gell, of his own oeuvre as an artist or agent, which would be a matter of aesthetics (Gell 1998: 237), but more generally a political intervention that collides colonial science and postcolonial art. A translation rather than just an interpretation. The camera becomes a second mask, returning Thompson's own gaze, making visible the visual thefts, the objectifications, to which Lévy-Bruhl referred. Thompson's photography works, thus, photologically – with the 'appearance' of knowledge.

*

XI. Archaeology as if photography

In the archive and in the field Archaeology and Photography fall in and out of step, revealing the ongoing nature of the photograph and the artefact as alternative temporal modalities of knowledge, of the archaeographic gesture as a way of living with the past. What they result in is not 'inflected with a predetermined overall meaning' that can be interpreted (Barthes 2013: 172). We should not exaggerate their finality, or their stillness, or their sheer materiality, or their indexicality, or their capacity adequately to be described through contemporary representation. The transformation of Visual Archaeology requires a sense of how Archaeology (including the rhythms of the public past) and Photography (including the pulse of private memory) live together in a scansion filled with intermittent convergences, stutters or hesitations or 'doubling-ups', unstable durational relations that effect transformation through 'an action particular to the Image' (Barthes 2011: 310, 421, note 8). The transformation of Visual Archaeology thus begins with acknowledging the transformative rather than purely representative dimensions of Archaeology and Photography. We need no longer reduce these methods to each other. Just as two-handedness brings creativity or binocular vision yields stereopsis through parallax, so through ongoing shifting patterns of bilateral discordance and concordance archaeography is a tool for refracting the photological dimensions of our knowledge of the world.

The void of the archaeological excavation is an aperture. Museum archives are dark rooms. Archaeography does not deal with ruins but transforms the world. Photology is visual knowledge of the past.

As technologies of duration Archaeology and Photography do not multiply moments of time but distribute modalities of knowledge. They fill the photological archive with disruptions and internal contradictions (Edwards 2001: 194–5, 236–7). As methods Archaeology and Photography interpret or read the past only in the sense that Barthes (2002) evokes of a kind of reading that changes what is read, not so much a simple mis-representation as the kind of active 'ruining' that Kate Briggs (2015) associates with translation, and that we see in the form of reverse translation in the photograph and the artefact as photological effects:

> And here I discover the method. I put myself in the position of one who does something, not one who speaks about something: I don't study a product, I assume a production; I abolish discourse on discourse; the world no longer comes to me in the form of an object, but through writing, that is to say through a practice: I proceed to another type of knowledge (that of the Amateur), and it is here that I am methodical. 'As If': isn't this formula the very expression of a scientific approach, as we see it in mathematics? I pose a hypothesis and I investigate, I discover the wealth of what comes from it.
>
> BARTHES 1984 [1978]: 325[36]

The photograph and the artefact are not straightforward traces, not indexes of 'what has been', not immutable mobiles, but intermittent and mutating effects, what 'appears to have been' (Barthes 2013: 71), and it follows that visual archaeology is not a form of representation but a transformative method. Knowledge made with light. This transformation of knowledge, this change of perspective, this figure-ground reversal (Wagner 1986), is the source of the political power of Homonymy, of Prefiguration, of the Unhistorical, of Impermanence, of Appearance.

Photology is Archaeology 'as if' Photography. A counter-speculative realism. As if a photograph were not a still. As if the past were more than a ruin. As if the archival box were a photographic aperture. As if the museum were a dark room; the vitrine a converging lens. As if artefacts were formed of the light cast from a blur of human days and nights. As if human memory could be seen. As if that photograph were me. As if archaeology could transform the world.

Notes

1   'La peinture n'est pas de la photographie, disent les peintres. Mais la photographie non plus n'est pas de la photographie, c'est-à-dire n'est pas de la copie'.

2   'À l'origine, le matériel photographique relevait des techniques de l'ébénisterie et de la mécanique de precision: les appareils, au fond, étaient des horloges à voir, et peut-être en moi, quelqu'un de très ancien entend encore dans l'appareil photographique le bruit vivant du bois'.

3   'en mécanisant à l'extrême le mode plastique du représentation'.

4 'Un image double – c'est à-dire la représentation d'un objet qui, sans la moindre modification figurative ou anatomique, soit même temps la representation d'un autre objet absolutement différent'.

5 For *mémoire involontaire* see Proust's *Swann's Way*. For *Optism-Unbewußten* see Benjamin 1977 [1931]: 371.

6 'Pourtant, cet hommage liminaire confinne l'existence du problème plutot qu'il ne le résout. La vraie réponse se trouve, croyons-nous, dans le charactère commun du mythe et de l'oeuvre musicale, d'être des langages qui transcendent, chacun à sa maniere, le plan du langage articulé, tout en requérant comme lui, et à l'opposé de la peinture, une dimension temporelle pour se manifester. Mais cette relation au temps est d'une nature assez particulière: tout se passe comme si la musique et la mythologie n'avaient besoin du temps que pour lui infliger un démenti. L'une et l'autre sont, en effet, des machines à supprimer le temps.'.

7 'ce moment très subtil où, à vrai dire, je ne suis ni un sujet ni un objet, mais plutôt un sujet qui se sent devenir objet: je vis alors une micro-expérience de la mort (de la parenthèse): je deviens vraiment spectre'.

8 'La Photographie, parfois, fait apparaître ce qu'on ne perçoit jamais d'un visage réel (ou réfléchi dans un miroir): un trait génétique, le morceau de soi-même ou d'un parent qui vient d'un ascendant'.

9 'se dévoile, se déchire, se révèle, au sens photographique du terme'.

10 *'L'exploitation de la rencontre fortuite de deux réalités distantes sur un plan non convenant'*.

11 'L'autre jour, j'ai relu le roman de Thomas Mann, *La Montagne magique*. Ce livre met en scène une maladie que j'ai bien connue, la tuberculose; par la lecture, je tenais rassemblés dans ma conscience trois moments de cette maladie : le moment de l'anecdote, qui se passe avant la guerre de 1914, le moment de ma proper maladie, alentour 1942, et le moment actuel, où ce mal, vaincu par la chimiothérapie, n'a plus du tout le même visage qu'autrefois. Or, la tuberculose que j'ai vécue est, a très peu de chose près, la tuberculose de *La Montagne magique*: les deux moments se confondaient, également éloignés de mon propre présent. Je me suis alors aperçu avec stupéfaction (seules les évidences peuvent stupéfier) que *mon propre corps était historique*. En un sens, mon corps est contemporain de Hans Castorp, le héros de *La Montagne magique*; mon corps, qui n'était pas encore né, avait déjà vingt ans en 1907, année où Hans pénétra et s'installa dans « le pays d'en haut », mon corps est bien plus vieux que moi, comme si nous gardions toujours l'âge des peurs sociales auxquelles, par le hasard de la vie, nous avons touché. Si donc je veux vivre, je dois oublier que mon corps est historique, je dois me jeter dans l'illusion que je suis contemporain des jeunes corps présents, et non de mon propre corps, passé.'.

12 See discussion by Blonsky (1985: xiii–xiv). The lecture was originally envisioned as *Proust et Moi* but eventually titled "Longtemps, je me suis couché de bonne heure".

13 'Photographs are valued because they give information. They tell one what there is; they make an inventory. To spies, meteorologists, coroners, archaeologists, and other information professionals, their value is inestimable. But in the situations in which most people use photographs, their value as information is of the same order as fiction' (Sontag 1973: 16).

14 'C'est ce que Michelet avait compris: l'Histoire, c'est en fin de compte l'histoire du lieu fantasmatique par excellence, à savoir le corps humain; c'est en partant de ce fantasme, lié chez lui à la résurrection lyrique des corps passés, que Michelet a pu faire de l'Histoire une immense anthropologie'.

15 'Des "systems d'instants" . . . se succédent, mais aussi se répondent. Car ce que le principe de vacillation désorganise, ce n'est pas l'intelligible du Temps, mais la logique illusoire de la biographie, en tant qu'elle suit traditionnellement l'ordre purement mathématique des années'.

16 Compare Candea (2016).

17 'un peu terrible qu'il y a dans toute photographie : le retour du mort'.

18 I am grateful to Louise Loe for this observation.

19 'For decades now historians have preferred to focus their attention on long periods of time, as if, beneath political shifts and their chapters, they undertook to bring to light the stable and almost indestructible system of checks and balances, the irreversible processes, the constant readjustments, the underlying tendencies that culminate and are reversed after centuries of continuity, the movements of accumulation and slow saturation, the great silent, immobile bases that the entanglement of traditional histories has covered over with the thickness of events. To carry out this analysis, historians have at their disposal tools that are partly of their own making and are partly inherited: models of economic growth, quantitative analysis of the flows of trade, profiles of demographic expansion and contraction, the study of climate and climate change, the identification of sociological constants, the description of technological adjustments, their diffusion and their and persistence. *These tools have enabled them to distinguish in the field of history various sedimentary strata; linear successions, which for so long had been the object of research, have given way to discoveries in depth.* From political movement to the slowness specific to "material civilization", the veins of analysis have multiplied: each has its own ruptures, each has a shape that belongs to it alone; and as we descend to the deepest layers, the rhythms become longer and longer.' (my emphasis). The original text reads: 'Voilà des dizaines d'années maintenant que l'attention des historiens s'est portée, de préférence, sur les longues périodes comme si, au-dessous des péripéties politiques et de leurs épisodes, ils entreprenaient de mettre au jour les équilibres stables et difficiles à rompre, les processus irréversibles, les regulations constantes, les phénomènes tendanciels qui culminant et s'inversent après des continuités séculaires, les mouvements d'accumulation· et les saturations lentes, les grands socles immobiles et muets que l'enchevêtrement des récits traditionnels avait recouverts de toute une épaisseur d'événements. Pour mener cette analyse, les historiens disposent d'instruments qu'ils ont pour une part façonnés, et pour une part reçus: modèles de la croissance économique, analyse quantitative des flux d'échanges, profils des développements et des régressions démographiques, étude du climat et de ses oscillations, repérage des constantes sociologiques, description des ajustements techniques, de leur diffusion et de leur persistance. Ces instruments leur ont permis de distinguer, dans le champ de l'histoire, des couches sédimentaires diverses; aux successions linéaires, qui avaient fait jusque-là l'objet de la recherche, s'est substitué un jeu de décrochages en profondeur. De la mobilité politique aux lenteurs propres à la "civilisation matérielle", les mveaux d'analyse se sont multipliés : chacun a ses ruptures spécifiques, chacun comporte un découpage qui n'appartient qu'à lui; et à mesure qu'on descend vers les socles les plus profonds, les scansions se font de plus en plus larges'.

20 'C'est la dernière aventure politique qu'aura connue le plus puissant pharaon de l'Egypte ancienne : Ramsès II qui avait vu l'exode des Hébreux, subi l'invasion des sauterelles et vaincu les farouches Hittites, est, aujourd'hui, l'enjeu d'une aigre polémique entre les savants anglo-saxons et français. La momie du grand roi avait été accueillie à l'aéroport du Bourget, il y a deux mois, avec les honneurs réservés aux chefs d'Etat : ministre, garde républicaine et fanfare. But de la visite : stopper la degradation du cadavre embaumé tel qu'on pouvait le voir au musée du Caire. « C'est une maladie diplomatique, insinuent certains égyptologues anglais et amèricains ; nos collègues français justifient par de prétendus dommages la venue du pharaon à Paris. » « Absurde, s'indigne le professeur Lionel Balout, administrateur du musée de l'Homme et responsable de l'opération Ramsès. L'odeur que dégageait la momie, au musée du Caire, prouverait à elle seule l'état de décompostion. Les bandelettes avaient été enlevées par les pillards des tombes royales et le corps était à peine protégé par des linges posés des siècles après l'embaumement. Nous y avons découvert des champignons microscopiques, des bactéries et même des larves d'origine récente. Un autre drame : des tensions se produisent entre les chairs desséchées par les résines et le squelette, ce qui provoque d'importantes fractures. » Pour arrêter cette évolution, on radiosérilisera le cadavre au cobalt 60. Les soins à Ramsès II doivent se terminer en mars. Les Egyptiens retrouveront alors leur souverain défunt, le premier qu'ils aient consenti à laisser partir pour une terre étrangère.'

21 The mistake is repeated many times in the subsequent literature – even in Paul Boghossian's (2007) discussion of Latour's paper in his highly conservative, ill-tempered and superficial critique of constructivist social science, *Fear of Knowledge*.

22 I am indebted to Daniel Antoine at the British Museum and Anne-Claire Salmas at the Griffith Institute, Oxford University for their assistance in understanding the facts of the medical archaeology of Ramesses II.

23 *'[L]a beauté n'a quelque fois qu'un instant*. La photographie saisit cette chance: elle montre, c'est son mot, l'instantané'.

24 'Les archives apportent donc autre chose: d'une part, elles constituent l'événement dans sa contingence radicale (puisque seule l'interprétation, qui n'en fait point partie, peut le fonder en raison); d'autre part, elles donnent une existence physique à l'histoire, car en elles seulement est surmontée la contradiction d'un passé révolu et d'un présent où il survit. Les archives sont l'être incarné de l'événementialité'.

25 'Le réalisme photographique ne distingue pas les accidents de la nature des choses: il les laisse sur le même plan'.

26 '"Was nie geschrieben wurde, lessen" heißt es bei Hofmannsthal. Der Leser, an den hier zu denken ist, ist der wahre Historiker'.

27 'Un immédiat antérieur. L'image s'accorde bien à ce leurre temporel: nette, surprise, encadrée, elle est déjà (encore, toujours) un souvenir (l'être de la photographie n'est pas de représenter, mais de remémorer)'.

28 'Une fois réalisé, celui-ci sera donc inévitablement décalé par rapport à l'intention initiale (d'ailleurs, simple schème), effet que les_surréalistes ont nommé avec bonheur « hasard objectif»'.

29 'Le punctum d'une photo, c'est ce hasard qui, en elle, me point (mais aussi me meurtrit, me poigne)'.

30 'Si fulgurant qu'il soit, le punctum a, plus ou moins virtuellement, une force d'expansion. Cette force est souvent métonymique.'.

31 'Comme âme vivante, je suis le contraire même de l'Histoire, ce qui la dément, la détruit au profit de ma seule histoire (impossible pour moi de croire aux "témoins"; impossible du moins d'en être un).'.

32 'Un masque n'est pas d'abord ce qu'il représente mais ce qu'il transforme, c'est à dire choisit de *ne pas* représenter. Comme un mythe, un masque nie autant qu'il affirme ; il n'est pas fait seulement de ce qu'il dit ou croit dire, mais de ce qu'il exclut.'.

33 'La manifestation politique fait voir ce qui n'avait pas de raisons d'être vu'.

34 Non seulement la Photo n'est jamais, en essence, un souvenir (dont l'expression grammaticale serait le parfait, alors que le temps de la Photo, c'est plutôt l'aoriste), mais encore elle le bloque, devient très vite un contre-souvenir'. For 'what appears to have been' and for more commentary on the tense, or 'temporal category' of photographs, see Barthes (2013: 71–4).

35 'Presque partout, en particulier, l'appareil photographique paraît spécialement dangereux. Les indigènes ignorants, dit M. Junod, ont une repulsion instinctive quand on veut les photographier. Ils disent : « Ces blancs vont nous voler et nous emporter au loin, dans des pays que nous ne connaissons pas, et nous resterons des êtres privés d'une partie de nous-mêmes. » Quand on leur montre la lanterne magique, on les entend qui plaignent les personnages représentés sur les images, et qui ajoutent : « Voilà ce qu'ils font de nous quand ils ont nos photographies ! » Avant que la guerre de 1894 éclatât, j'étais allé montrer la lanterne magique dans des villages païens éloignés. Les gens m'accusèrent d'avoir causé ce malheur, en faisant ressusciter des hommes morts depuis longtemps'.

36 'Et je retrouve ici, pour finir, la méthode. Je me mets en effet dans la position de celui qui *fait* quelque chose, et non plus de celui qui parle *sur* quelque chose: je n'étudie pas un produit, j'endosse une production; j'abolis le discours sur le discours ; le monde ne vient plus à moi sous la forme d'un objet, mais sous celle d'une écriture, c'est-à-dire d'une pratique: je passe à un autre type de savoir (celui de l'Amateur) et c'est en cela que je suis méthodique. "Comme si": cette formule n'est-elle pas l'expression même d'une démarche scientifique, comme on le voit en mathématiques? Je fais une hypothèse et j'explore, je découvre la richesse de ce qui en découle'.

References

Adams, J.E. 2016. Scientific Studies of Pharaonic Remains: Imaging. In C. Price, R. Forshaw, A. Chamberlain and P. Nicholson (eds) *Mummies, Magic and Medicine in Ancient Egypt: Multidisciplinary Essays for Rosalie David.* Manchester: Manchester University Press, pp. 371–386.

Azoulay, A. 2012. *Civil Imagination: A Political Ontology of Photography.* London: Verso.

Barthes, R. 1961. Le message photographique. *Communications* 1: 127–138.

Barthes, R. 1977. *Fragments d'un discours amoureux.* Paris, Seuil.

Barthes, R. 1978. *Leçon. Texte de la leçon inaugurale prononcée le 7 janvier 1977 au Collège de France.* Paris: Editions du Seuil (published in English, translated by R. Howard, in 1979 as Lecture in Inauguration of the Chair of Literary Semiology, Collège de France, January 7 1977 *October* 8: 3–16.)

Barthes, R. 1980. *La Chambre claire.* Paris, Seuil/Gallimard.

Barthes, R. 1984 [1978]. Longtemps, je me suis couché de bonne heure. In *Le Bruissement de la Langue (Essais critiques IV)*. Paris: Editions du Seuil, pp. 313–325.

Barthes, R. 2002. *Comment vivre ensemble: simulations romanesques de quelques espaces quotidiens: notes de cours et de séminaires au Collège de France, 1976–1977* (edited by C. Coste). Paris: Editions du Seuil.

Barthes, R. 2003. *La préparation du roman, I et II: cours et séminaires au Collège de France, 1978–1979 et 1979–1980* (edited by N. Léger). Paris: Editions du Seuil.

Barthes, R. 2011. *The Preparation of the Novel: lecture courses and seminars at the Collège de France, 1978–1979 and 1979–1980* (trans. K. Briggs). New York: Columbia University Press (translation of Barthes 2003).

Barthes, R. 2013. *How to Live Together: Novelistic Simulations of Some Everyday Spaces* (trans. K. Briggs). New York: Columbia University Press (translation of Barthes 2002).

Bazin, A. 1960. The Ontology of the Photographic Image (trans. H. Gray). *Film Quarterly* 13(4): 4–9.

Benjamin, W. 1974 [1940]. Paralipomena zu den Thesen über den Begriff der Geschichte. In *Gesammelte Schriften Volume 2*. Frankfurt: Suhrkamp, pp. 1228–1246.

Benjamin, W. 1977 [1931]. Kleine Geschichte der Photographie. In *Gesammelte Schriften Volume 2*. Frankfurt: Suhrkamp, pp. 368–385.

Berger, J. 1972. *Ways of Seeing*. London: Penguin.

Berger, J. 1976. Drawn to That Moment. *New Society* 37: 81–82.

Berger, J. and J. Mohr 2016 [1982]. *Another Way of Telling: A Possible Theory of Photography*. London: Bloomsbury.

Berliner, D. 2014. On Exonostalgia. *Anthropological Theory* 14(4): 373–386.

Bloch, E. 1959. *Werkausgabe: Band 5: Das Prinzip Hoffnung*. Suhrkamp, Frankfurt am Main.

Blonsky, M. 1985. Introduction: The Agony of Semiotics: Reassessing the Discipline. In M. Blonsky (ed.) *On Signs*. Baltimore, MD: Johns Hopkins University Press, pp. xiii–li.

Bloor, D. 1999. Anti-Latour. *Studies in the History and Philosophy of Science* 30(1): 81–112.

Boghossian, P. 2007. *Fear of Knowledge: Against Relativism and Constructivism*. Oxford: Oxford University Press.

Bourdieu, P. 1990. The Cult of Unity and Cultivated Differences. In P. Bourdieu, L. Boltanski, R. Castel, J-C. Chamboredon and D. Schnapper *Photography: A Middle-brow Art* (trans. S. Whiteside). London: Polity, pp. 13–72.

Breton, A. 1928. *Nadja*. Paris: Gallimard.

Breton, A. 1969 [1935]. The Surrealist Situation of the Object. In *Manifestoes of Surrealism* (trans. R. Seaver and H.R. Lane.). Ann Arbor: University of Michigan Press, pp. 255–278.

Briggs, K. 2015. Practising with Roland Barthes. *L'Esprit Créateur* 55(4): 118–130.

Briggs, K. 2017. *This Little Art*. London: Fitzcaraldo Editions.

Candea, M. 2016. De deux modalités de comparaison en anthropologie sociale. *L'homme* 218: 183–218.

Cardinal, R. 1986. *Nadja*. London: Grant and Cutler.

Caws, M.A. 1988. Linkings and Reflections: André Breton and his Communicating Vessels. *Dada/Surrealism* 17: 91–100.

Clarke, D. 1973. Archaeology: The Loss of Innocence. *Antiquity* 47: 6–18.

Crevel, R. 1925. Le miroir aux objets. *L'art vivant* 14 (July 15 1925): 23–24.

Dali, S. 1930. L'ane pourri. *Le Surréalisme au service de la revolution* 1: 9–12.

Daston, L. 2008. On Scientific Observation. *Isis* 99(1): 97–110.

Daston, L. 2017. Third Nature. In L. Daston (ed.) *Science in the Archives: Pasts, Presents, Futures.* Chicago: University of Chicago Press, pp. 1–14.

Daston, L. and P. Galison 1992. The Image of Objectivity. *Representations* 40: 81–128.

de Duve, T. 1978. Time Exposure and Snapshot: The Photograph as Paradox. *October* 5: 113–125.

Deleuze, G. 2004 [1967]. How Do We Recognise Structuralism? In *Desert Islands and Other Texts, 1953–1974.* Los Angeles: Semiotexte, pp.170–192.

Desroches-Noblecourt, C. 1996. *Ramses II – la veritable histoire.* Paris: Pygmalion/G. Watelet.

Didi-Huberman, G. 2002. The Surviving Image: Aby Warburg and Tylorian Anthropology. *Oxford Art Journal* 25(1): 61–69.

Dubois, P. 2016. Trace-Image to Fiction-Image: The Unfolding of Theories of Photography from the '80s to the Present. *October* 158: 155–166.

Duchamp, M. 1934. *La boîte verte. La mariée mise à nu par ses célibataires même.* Paris: Edition Rroce Sélavy.

Edwards, E. 2001. *Raw Histories: Photographs, Anthropology and Museums.* Oxford: Berg.

Ernst, M. 1970 [1936]. Au-delà de la peinture. In *Écritures.* Paris: Gallimard, pp. 252–256.

Forestier J. and J. Rotes-Querol 1950. Senile Ankylosing Hyperostosis of the Spine. *Annals of the Rheumatic Diseases* 9: 321–330.

Foucault, M. 1969. *L'Archéologie du Savoir.* Paris: Gallimard.

Fox Talbot, W.H. 1844. *The Pencil of Nature.* London: Longman, Brown, Green and Longmans.

Gell, A. 1998. *Art and Agency: An Anthropological Theory.* Oxford: Clarendon.

Genette, G. 1980. *Narrative Discourse: An Essay in Method* (trans. J.E. Lewin). New York: Cornell University Press.

Gilchrist, S. 2016. Everywhen: The Eternal Present in Indigenous Art from Australia. In S. Gilchrist (ed.) *Everywhen: The Eternal Present in Indigenous Art from Australia.* Cambridge, MA: Harvard Art Museums, pp. 18–31.

Harris, H.A. 1926. The Growth of the Long Bones in Childhood with Special Reference to Certain Bony Striations of the Metaphysis and the Role of Vitamins. *Archives of Internal Medicine* 38: 785–793.

Harris, S. 2001. The Return of the Dead: Memory and Photography in W.G. Sebald's 'Die Ausgewanderten'. *German Quarterly* 74(4): 379–391.

Hawass, Z.A. and S. Saleem 2016. *Scanning the Pharaohs: CT Imaging of the New Kingdom Royal Mummies.* Cairo: American University in Cairo Press.

Henderson, J. 1869. Photography as an Aid to Archaeology. *The Photographer's Journal* 13: 37–39.

Hicks, D. 2003. Archaeology Unfolding: Diversity and the Loss of Isolation. *Oxford Journal of Archaeology* 22(3): 315–329.

Hicks, D. 2010. The Material-cultural Turn: Event and Effect. In D. Hicks and M.C. Beaudry (eds) *The Oxford Handbook of Material Culture Studies.* Oxford: Oxford University Press, pp. 25–98.

Hicks, D. 2016. The Temporality of the Landscape Revisited. *Norwegian Archaeological Review* 49(1): 5–22.

Hicks, D. and S. Mallet 2019. *Lande: the Calais 'Jungle' and Beyond.* Bristol: Bristol University Press.

Hornby, L. 2017. *Still Modernism: Photography, Literature, Film.* Oxford: Oxford University Press.

Ingold, T. 2013. *Making: Anthropology, Archaeology, Art and Architecture.* Abingdon: Routledge.

Kelsey, R. 2015. *Photography and the Art of Chance.* Cambridge, MA: Harvard University Press.

Kracauer, S. 1960. *Theory of Film: The Redemption of Physical Reality.* Oxford: Oxford University Press.

Kracauer, S. 1993. Photography (trans. T.Y. Levin). *Critical Inquiry* 19(3): 421–436.

Kriebel, S.T. 2007. Theories of Photography: A Short History. In J. Elkins (ed.) *Photography Theory.* London: Routledge, pp. 3–49.

Lane Fox, A.H. 1875. On the Evolution of Culture. *Journal of the Royal Institution* 7: 357–389 In Anon (ed.) *Notices of the Proceedings at the meetings of members of the Royal Institution of Great Britain with abstracts of the discourses delivered at evening meetings, Volume 7 (1873–1875).* London: William Clowes and Sons, pp. 496–520.

Latour, B. 1999a. *Pandora's Hope: Essays on the Reality of science studies.* Cambridge, MA: Harvard University Press.

Latour, B. 1999b. On the Partial Existence of Existing and Nonexisting Objects. In L. Daston (ed.) *Biographies of Scientific Objects.* Chicago: University of Chicago Press, pp. 247–269.

Latour, B. 2007. A Textbook Case Revisited. Knowledge as Mode of Existence. In E.J. Hackett, M. Lynch, J. Wajcman and O. Amsterdamska (eds). *Handbook of Science and Technology Studies* (third edition). Cambridge: MIT Press, pp. 83–11.

Lydon, J. 2014. Introduction: The Photographic Encounter. In J. Lydon (ed.) *Calling the Shots: Indigenous Photographies.* Canberra: Aboriginal Studies Press, pp. 1–20.

Lydon, J. 2016. Transmuting Australian Aboriginal Photographs. *World Art* 6(1): 45–60.

Lévi-Strauss, C. 1962. *La Pensée Sauvage.* Paris: Librarie Plon.

Lévi-Strauss, C. 1964. *Mythologiques: Le Cru et le Cuit.* Paris: Librarie Plon.

Lévi-Strauss, C. 1997 [1993]. *Look, Listen, Read* (trans. B.C.J. Singer). New York: Basic Books.

Lévi-Strauss, C. 2008 [1979]. *La Voie des Masques.* In *Oeuvres.* Paris: Editions Gallimard, pp. 875–1050.

Lévy-Bruhl, L. 1922. *La mentalité primitive.* Paris: Presses universitaires de France

Mirzoeff, N. 2017. *The Appearance of Black Lives Matter.* Miami: Name Publications. https://namepublications.org/item/2017/the-appearance-of-black-lives-matter/

Petrie, W.M.F. and F.L. Griffith 1898. *Deshasheh, 1897: Fifteenth Memoir of the Egypt Exploration Fund.* London: Egypt Exploration Fund.

Pinney, C. 2003. How the Other Half . . . In C. Pinney and N. Peterson (eds) *Photography's Other Histories.* Durham, NC: Duke University Press, pp. 1–14.

Pinney, C. 2005. Things Happen: Or, from Which Moment Does That Object Come? In D. Miller (ed.) *Materiality.* Durham, NC: Duke University Press, pp. 256–272.

Pinney, C. 2010. Camerawork as Technical Practice in Colonial India. In T. Bennett and P. Joyce (eds) *Material Powers: Cultural Studies, History and the Material Turn.* London: Routledge, pp. 145–170.

Pinney, C. 2012. Seven Theses on Photography. *Thesis Eleven* 113(1): 141–156.

Rancière, J. 1997. Onze thèses sur la politique. *Filozofski vestnik* 18(2): 91–106.

Richter, G. 2010. Introduction. Between Translation and Invention: The Photograph in Deconstruction. In J. Derrida *Copy, Archive, Signature: A Conversation on Photography* (trans. J. Fort). Stanford, CA: Stanford University Press, pp. ix–xxxviii.

Schiffer, M.B. 1975. Archaeology as Behavioral Science. *American Anthropologist* 77(4): 836–848.

Shepherd, N. 2015. *Mirror in the Ground: Archaeology, Photography and the Making of a Disciplinary Archive.* Johannesburg: Jonathan Ball Publishers.

Silverman, K. 2015. *The Miracle of Analogy or The History of Photography, Part 1.* Stanford, CA: Stanford University Press.

Sontag, S. 1973. *On Photography.* New York: Farrar, Straus and Giroux.

Sterling, C. 2015. Rethinking Heritage and Photography: Comparative Case Studies from Cyprus and Cambodia. Unpublished Ph.D. thesis, University College London.

Sterling, C. 2017. Mundane Myths: Heritage and the Politics of the Photographic Cliché. *Public Archaeology* 15(2–3): 87–112.

Strathern, M. 2005. *Kinship, Law and the Unexpected: Relatives Are Always a Surprise.* Cambridge: Cambridge University Press.

Strathern, M. 1999. *Property, Substance and Effect: Anthropological Essays on Persons and Things.* London: Athlone Press.

Tardrew, C. 1976. Nos savants au secours de Ramses, tombe malade 3000 ans après sa mort. *Paris Match* 1435 (26 November 1976): 74–75.

Thompson, C. 2012. Artist Statement. In C. Morton (ed.) *Spiritual Repatriation and the Archive.* Oxford: Pitt Rivers Museum.

Tylor, E.B. 1881. *Anthropology.* London: Macmillan.

Viveiros de Castro, E. 2004. Perspectival Anthropology and the Method of Controlled Equivocation. *Tipití: Journal of the Society for the Anthropology of Lowland South America*: 2(1): 3–22.

Wagner, R. 1986. *Symbols That Stand for Themselves.* Chicago: Chicago University Press.

Wilson, D.W. 1982. *Air Photo Interpretation for Archaeologists.* Cambridge: Cambridge University Press.

INDEX

Printed and bound by CPI Group (UK) Ltd, Croydon, CR0 4YY

17/10/2024

01775682-0012